Transport
and Development
in the Third World

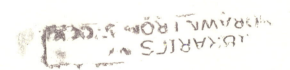

ROUTLEDGE

London and New York

First published 1996
by Routledge
11 New Fetter Lane, London, EC4P 4EE

Transferred to Digital Printing 2004

Simultaneously published in the USA and Canada
by Routledge
29 West 35th Street, New York, NY 10001

Typeset in Times by J&L Composition Ltd, Filey, North Yorkshire

British Library Cataloguing in Publication Data
A catalogue record for this book is available from the British Library

Library of Congress Cataloguing in Publication Data
Simon, David
    Transport and development in the Third World/David Simon
        p.   cm. – (Routledge introductions to development)
    Includes bibliographical references and index.
        1. Transportation–Developing countries–Case studies.
    2. Developing countries–Economic conditions–Case studies.
    I. Title.   II. Series.
    HE148.5.S55   1996
    388′.09172′4–dc20   95–40958

ISBN 0–415–11905–7

Contents

It
de
a
p
a
ut
th
po
qu
st
iss

Da
Un
Re

Plates

Figures

Tables

Preface

The diversity of forms of transport, not to mention the arrangements and procedures surrounding their use, found around the Third World and beyond is so great that any attempt to encapsulate them within the scope of a short introductory textbook such as this is inevitably fraught with difficulty. It is impossible to include everything of relevance, to cite examples from every country and region or to indicate the full extent of variations which exist on any particular topic. I have therefore necessarily had to be selective in coverage and have organised the material thematically (rather than by mode of transport or the purpose of movement) in an attempt to develop coherent and, I hope, consistent and stimulating arguments. These have been illustrated with examples selected for their interest, the availability of suitable documentation and my familiarity with them. I make no claim for total comprehensiveness and am well aware that many other examples or case studies might have been used instead.

Perhaps the greatest risk in an undertaking of this kind is that of generalisation across large numbers of countries and situations where conditions do vary considerably. Ironically, the use of a small number of categories or labels in attempting to make sense of the wide diversity which does exist and to highlight and explain certain significant differences between the groups makes this almost inevitable. However, I have sought to indicate the scope for generalisation in terms of underlying processes and causal relationships rather than just empirical detail.

The objectives of this book are to introduce readers to some of the pertinent transport issues in Third World countries and to highlight how these relate to wider development problems and processes. I have thus deliberately sought to be provocative by challenging the conventional approaches and practices within transport policy and planning, which I argue to be unfortunately narrow and inappropriate in both theoretical and practical terms. Not everyone will agree, but if readers are at least prompted to reflect critically on some of the key issues addressed, then the book will have succeeded in its aims.

The material covered here has been collected and refined over a number of years in the context of field research and consultancy work, the teaching of transport issues and supervision of dissertations at both undergraduate and postgraduate levels. I therefore owe a debt of gratitude to many colleagues and past and current students, particularly Mongezi Noah and Keith Etherington. David Drakakis-Smith eventually convinced me that I should write this book, while Tristan Palmer and Matthew Smith at Routledge have been efficient and encouraging throughout. Jeff Turner of the Overseas Centre, Transport Research Laboratory (TRL), provided copies of recent relevant TRL output and willingly checked many details. Particular acknowledgement is also due to Sheryl and Jonathan, my wife and son, for their tolerance, support and the necessary distractions. Justin Jacyno drew the figures, Caroline Stafford typed the tables while Keith Etherington and David Hilling generously provided Plate J.1 and Plates 6.2, 6.3, 6.4, 6.5 and 6.6 respectively. All other plates were provided by the author.

David Simon
June 1995

1
Introduction

Aims and scope: transport and development

It would not be an exaggeration to describe transport as something of a Cinderella within Development Studies, being of fundamental importance yet little appreciated and often regarded within the social sciences as 'unimaginative' or 'technical'. The reasons for this are no doubt numerous but two stand out. First, many people are familiar with the often eccentric obsession with trains, planes, cars or ships on the part of some transport enthusiasts, making it difficult to take the subject of their infatuation seriously. Second, and no less important, is the widely perceived lack of conceptual or explicit critical theoretical content in mainstream Transport Studies. Such concerns are, frankly, often justified.

The situation can be explained in terms of two related factors, namely the heavy influence of transport or traffic engineering, a discipline adhering to an ethos of 'rational science' and concerned primarily with technical issues of project design and implementation, and, second, the unfortunate way in which transport has been taught over recent decades within the social sciences. The subdiscipline has been deeply immersed in a rather uncritical tradition within conventional neoclassical economics, where the crucial assumptions and axioms have too often been implicit rather than explicit. People arguing for alternative perspectives have generally been relegated to the margins or dismissed as idealists and cranks. Transport Studies and Transport Geography have been slower than most other social sciences to adapt to change or to adopt more critical perspectives, more challenging theories and to overcome the Eurocentric bias of 'positive' (i.e. supposedly value-free) neoclassical economics, in terms of which Northern values, theories and methods are simply assumed to have universal applicability.

In other words, not only are many of the underlying assumptions of this conventional economic philosophy unrealistic but they are applied unproblematically by Western economists to situations around the world as if they were equally relevant and appropriate in all socio-cultural, political and economic contexts. This is often not the case, and the way in which international agency staff and private transport and economic consultants generally operate, applying Western analytical methods and prescribing or recommending narrowly conceived Western solutions, is itself problematic. Indeed, it illustrates quite profoundly the necessity for a broader, more critical, conception of transport and its role as a complement to conventional technical or 'instrumental' expertise.

A similar limitation characterises most of the published transport literature, including textbooks. For example, in *Transportation and World Development*, one of the more recent and widely used texts in the subject, Wilfred Owen provides a narrow descriptive overview of changing transport technologies and different mode/vehicle use patterns around the world. The following example is typical:

> Comparing the situation in the United States with conditions in India illustrates the size of the transport discrepancy. If everyone in the United States decided to travel by automobile at one time, there would be 534 cars available for every 1,000 persons, and no one would need to ride in the back seat. Every 740 Indians would have to share one car. There are more trucks in Kentucky (population 3.6 million) than in all of India (population 750 million).
>
> (ibid.: 9–11)

Such figures are, of course, both interesting and relevant (see also Table 1.1). However, the argument is taken no further; he makes no real attempt to explain or even discuss why there are such major differences or how they have arisen. The only framework offered is a short and descriptive summary of 'The Stages of Mobility' (ibid.: 3–7), which suggests that all societies and countries have passed or will pass through five successive stages in an unproblematic and linear manner. The author's use of the form ('The Stages . . .') implies a single development path. The possibility of different experiences, or the bypassing of particular stages, is not addressed or even implied. The author's tone

Table 1.1 Travel and freight mobility indices relative to GNP per capita

Country	GNP per capita	Travel mobility	Freight mobility
Switzerland	139	104	81
Sweden	119	96	151
Federal Republic of Germany	117	101	57
Belgium	109	88	117
Norway	106	55	107
United States	106	160	260
Netherlands	101	83	42
France	100	100	100
Canada	95	114	374
Australia	91	107	335
Japan	87	96	94
United Kingdom	63	78	47
Czechoslovakia	53	54	132
Italy	53	86	49
Spain	43	54	44
USSR	40	34	229
Hungary	38	34	68
Venezuela	31	24	36
Yugoslavia	24	32	55
Argentina	24	32	114
Iran	32	10	10
Brazil	18	18	23
Mexico	15	14	42
Korea, Republic of	15	8	16
Malaysia	14	11	26
Turkey	14	5	26
Ecuador	11	5	21
Colombia	11	6	47
Nigeria	6	5	5
Philippines	6	2	18
Egypt	5	5	13
Indonesia	3	3	5
Pakistan	2	3	10
China	2	3	16
India	2	5	26
Ethiopia	1	2	3
Bangladesh	1	2	3

Note: Data on a per capita basis include estimated passenger-miles per year by automobile, rail and domestic air travel; miles of railway, ton miles of railway freight, miles of roads and numbers of trucks
Source: W. Owen (1987: 10)

throughout is rather deterministic and Northern-centric. According to this view, a major characteristic of poor countries is that they have inadequate and often rudimentary transport systems; that they have a lot of catching up to do and, by implication, should abandon traditional modes and vehicles in the process. Thus we are told variously that:

> The transport symbols of traditional societies are dirt tracks, bullock carts, camels, crude country boats, and human beings toiling as beasts of burden. The United States itself was subject to these same conditions not so many years ago.
>
> (ibid.: 4)

> Many less developed areas have been largely untouched by the transport revolution.
>
> (ibid.: 7)

> The size of the mobility gap between the rich countries and the poor ones, and the relation between mobility and economic progress, can be seen in the comparison between gross national product per capita and transport facilities and traffic.
>
> (ibid.: 8)

> Immobility and poverty go hand in hand.
>
> (ibid.: 9)

> Reducing the lag in Third World development is a complex and long-term problem, which calls for overcoming isolation in order to supply capital, technology, education, and the necessary innovations in policy, organization, and management.
>
> (ibid.: 142)

Fortunately, Owen is aware of the fallacy of one potential logical conclusion that could be drawn from such perspectives, and explicitly warns against it, even within the confines of purely economic development:

> Comparisons of income and mobility are not meant to imply that transportation by itself is capable of achieving economic development. It is a necessary but not sufficient element in the development process, and many costly transport undertakings have turned out to be extremely wasteful of resources because they were not accompanied by other actions to further economic progress.
>
> (ibid.: 11)

This is perfectly correct and important to note; transport alone will not 'solve' development problems. However, different countries face different problems, which have arisen in different ways. Solutions are likely to differ too, especially if they are to be locally appropriate. This also requires an understanding of underlying causes and how particular problems have arisen. By contrast, Owen's vision is one of a single development future, characterised by transport uniformity, global homogeneity and human progress. He ends his book in overly idealistic tones and almost triumphalist zeal:

> In summary, the current stage in the evolution of transportation introduces the possibility that global mobility can bring about a new era of international collaboration aimed at greater efficiency and equity in the use of the world's resources. The new mobility makes possible a common vision. . . . Under these circumstances, transportation policy makers and the transportation industries are offered an exceptional challenge and opportunity. By helping to overcome the immobility and isolation that contribute to economic underachievement, they can set the stage for increased world output, expanded international trade, higher per capita incomes, and improved levels of living for the majority of humankind. As the transportation tasks that lie ahead continue to assume global dimensions, those involved in the age-old struggle to overcome time and distance are suddenly called upon to serve two countries – their own and the planet earth.
>
> (ibid.: 142)

Again, powerful forces for globalisation are certainly operating, and transport and communications systems are becoming increasingly global in scope if not universal in reach or quality. However, Owen's vision of homogeneity, peaceful progress and lack of conflict seems Utopian. The very same processes fostering globalisation in certain spheres and transport modes are heightening contradictions and conflicts in others: between groups of people, communities in different localities, different transport priorities and competing resource uses. Progress and development may mean very different things to different people and in different contexts, while conflicts are showing no signs of lessening as technologies and means of communication become more sophisticated. Indeed, the opposite may be occurring. Moreover, the bullock cart or rickshaw may represent as much a symbol of the way forward as part of an integrated and appropriate transport system, as motorised, 'modern' modes of transport which have a high import component, generate pollution and often represent inefficient use of roadspace and other resources.

Elaboration of these arguments forms the central thesis of this book. By broadening the scope of coverage, the objective is not only to stimulate the interest of students but also to demonstrate that transport issues are both important and relevant within Development Studies. My concern is therefore with transport *and* development, or more precisely, transport *in* development.

Naturally, the book surveys prevailing transport conditions and current policy concerns in countries of the South or Third World, highlighting the differences among countries but also among different areas and groups of people within them. However, these descriptive elements are complemented by more analytical material which addresses such questions as: What are the relationships between investment in different forms of transport infrastructure and development? Can improved transport links and accessibility actually have an adverse impact on local or national development? How are rapidly growing metropolises in the South attempting to deal more effectively with their increasingly chronic traffic congestion and associated pollution? What are the likely ethical, politico-economic and social implications of the introduction of new, 'hi-tech' traffic management tools, transport modes like mass rapid transits, or fashionable policies like transport liberalisation and privatisation in non-Western societies?

By providing accessible surveys of different theoretical approaches and current issues and policies, illustrated with examples drawn from a wide variety of contexts in different countries, the book is designed to introduce students to the subject in a topical and thematic manner. Connections and interrelationships between different disciplines, economic sectors, localities and regions are highlighted in order to locate transport – as a field of research and study, a sphere of practical importance and terrain of conflict – more fully within the sphere of Development Studies and wider social-scientific enquiry.

Basic terms and concepts

It is important to clarify what is meant by several key terms and concepts at the outset.

Transport

Most people have a clear idea of the respective *modes* of transport (road, rail, water, air, etc.) and some of the different forms of each (e.g. lorry, car, bus, autorickshaw, bicycle or handcart in the case of road transport) based on their own travel experiences. However, in this book the term is used to denote these modes of transport as well as the associated infrastructure and institutional frameworks, arrangements and policies, issues of access to and use of transport by different groups, and the causes and implications of such considerations. It is important to emphasise this broader, more holistic focus rather than being concerned purely with vehicles, routes and their use and regulation.

Development

I understand 'development' to represent a broad, multifaceted process whereby quality of life and a sense of fulfilment are enhanced. It can apply to an individual as much as to a group or country. Naturally, this includes many qualitative as well as quantitative aspects and is therefore difficult to measure directly as a whole. Development is also a long-term process, so that the impact of interventions or changes may take a considerable period to become visible or to work through the system fully. Perhaps as a consequence, there has been a persistent tendency to focus on the material, more readily quantifiable dimensions. In the era of decolonisation following the Second World War, 'development' was generally and unproblematically regarded as being synonymous with economic development, which in turn was measured by the ratio of a country's Gross National Product to its population (i.e. GNP per capita).

Since at least the 1970s, the importance of gender, social, cultural, political and environmental dimensions of development have been widely recognised and efforts have been made to find a more balanced and appropriate form of measurement, such as the composite Physical Quality of Life Index (PQLI). This comprises three variables, namely life expectancy, infant mortality and adult literacy, with each country represented on a scale of 0–100. One of the most recent and widely discussed composite indices is the Human Development Index (HDI), first published by the United Nations Development Programme (UNDP) in 1990. This bears some similarity to the PQLI and comprises three elements: the real purchasing power of GNP per capita, as an indicator of material standards of living; the average level of schooling, as a proxy for educational and cultural opportunities; and average life expectancy at birth, as an indicator of the general level of access to health facilities. Countries are rated on a scale of 0–1, with 0–0.499 representing low levels of human development, 0.5–0.799 medium levels, and 0.8–1.0 high levels of human development respectively. As with the PQLI, this produces a rather different ranking of countries from that based on GNP per capita alone, giving considerable weight to social expenditure and access to social resources. Inevitably, the HDI has received some criticism despite a generally favourable reception, and is still being refined and improved. The latest details can be found in the UNDP's annual *Human Development Report*.

It is interesting to note that composite development indices generally do not include a mobility element, despite the widely claimed importance of transport and human interaction over space to any development process. Owen (1987: 8–11) provides one of the few attempts to explore a systematic relationship between mobility and GNP per capita. His mobility index combines data from a range of sources on transport facilities as well as the movement of both passengers and freight. Details of its construction are not given, but it yields a relative ranking on three variables, namely GNP per capita, travel mobility and freight mobility. The value of France on each variable is set to 100 as the benchmark. Although Owen is working very much within the modernisation paradigm and concerned only with economic development, as explained above, necessitating great caution against imputing causal explanations to such statistical relationships, the data are interesting at a descriptive level (Table 1.1). They provide a single snapshot for the early 1980s; ideally the exercise should be undertaken at regular intervals to give an indication of the direction and extent of change. Unfortunately, Owen fails to mention either the likelihood of change over time or that it could be quite rapid in certain countries; he uses the table merely to indicate 'the broad dimensions of the transportation gap' (ibid.: 9).

Notwithstanding recent innovations like the HDI, designed to provide more comprehensive representations of development, disproportionate attention is still frequently devoted by development planners and practitioners to the economic sphere, which they hold to be the 'engine' of development. At the extreme, there is still a widespread but erroneous view that if economic development can be promoted, everything else will somehow follow more or less spontaneously. Given the dominance of engineers and economists in Transport Studies, for example, it is understandable that such views and the paradigms which underpin them have remained dominant. However, development is categorically *not* reducible to economic development. Even the World Bank, which is by far the world's largest and most influential international development agency, is regularly criticised for continuing to regard economic impacts as the primary yardstick, despite undoubtedly having made considerable progress in enhancing the importance attached to other aspects of development.

GNP per capita also remains the Bank's principal variable for classifying countries. In all the tables in the Bank's annual *World Development Report*, countries are ranked from lowest to highest per capita GNP and divided into categories for aid policy purposes on that basis. In the 1994 Report, low-income countries are defined as having a per capita GNP of US$675 or less in 1992; lower middle-income countries lie in the range of more than $675 up to $2,695, upper middle-income countries lie between $2,695 and $8,355, and high-income countries lie above that threshold (World Bank 1994: x).

The growing concern with environmental issues and the now widely accepted interdependence of development–environment processes, has forced major development agencies and banks to accord greater weight to the environmental impacts of development projects and programmes, including those within the transport sector. Project documents at least pay lip service to the desirability of sustainable development, even though no dramatic or consistent change in the basic nature of projects being undertaken is yet occurring. Despite some recent attempts to include the environmental costs of development in national accounts through variations of so-called 'natural resource accounting' which involve reducing GNP through applying monetary values to pollution levels, deforestation and the like, no widely used composite development index yet includes an environmental variable.

Third World and the South

The term 'Third World' has remained controversial since it was first coined in the mid-1950s. While some have regarded it as demeaning, relegating the countries thus classified to 'third-class' status, others – including many political leaders in those countries – have used it as a rallying banner for collective action, rather like the 'third estate' in revolutionary France, to press for a New International Economic Order. More recent concerns about the term have centred on the increasing diversity of countries thus classified (arguing that further subdivisions are necessary), and also on how meaningful it can be to continue referring to a Third World now that the erstwhile Second World of centrally planned economies no longer exists. While such arguments are undoubtedly important, it is certainly true that the term remains in widespread use and is generally understood.

I use the term 'Third World' here in a historically rooted sense to indicate those countries which did not become industrialised and wealthy during the process of establishing the existing world order. Most, but not all, of these countries suffered European colonisation and exploitation. Today, the most problematic countries in terms of this classification are some of the Newly Industrialising Countries, especially Taiwan, Singapore and increasingly also Hong Kong, which have per capita incomes on a par with several West European countries and are highly industrialised. Nevertheless, historically they fitted the category well. In this book, 'Third World' is used interchangeably with 'the South', a term coined by the Brand Commission in 1980 to designate the poorer countries in contrast to the wealthier and powerful countries of 'the North'. Of course, many countries of the South are not geographically located in the Southern Hemisphere.

Pattern versus process

Although quite distinct, these two terms are frequently confused or used synonymously. A *pattern* (or profile) is generally a static picture or snapshot of one or more situations at a given time. For example, one can examine vehicle use patterns in one or more specific places or among one or more given groups of people. When used to denote the passage of time, as in a behaviour pattern, the term implies a regularity or consistency in such behaviour.

By contrast, a *process* denotes a dynamic trend, movement or change over time, as with 'development' in the sense defined above. The term 'process' itself says nothing about the pace or direction of change, which could be consistent or variable. In real life, changes generally occur at different intervals and rates, and have varying impacts, rather than being totally regular. For example, the introduction of a new transport technology into a particular locality could be ignored, rejected or adopted. If adopted, it could take place across the board immediately; at a uniform rate; or rapidly at first and then tapering off as the potential market becomes saturated. The outcome depends on specific local conditions and the perceived relevance, value or appropriateness of the innovation. Processes can be studied either through ongoing monitoring or by means of comparative static research, in other words a series of snapshots at different (and ideally) significant moments during that process.

Description versus analysis

These are two different levels of study or research. *Description* is more basic or superficial, being concerned purely with describing or summarising data or other information, together with any regularities in those observations. For example, descriptive statistics provide a series of simple measures about a dataset, such as the average or mean value, the range of observed values, the relative or absolute frequency with which each number occurs, the variance and standard deviation from the mean of values in the set.

Analysis, on the other hand, involves seeking to explain the observed data or behaviour, i.e. attempting to find causes and interrelationships underlying and explaining the observations. Again, this can be undertaken within broader or narrower confines, depending on the assumptions made or the range of one's concerns. Understanding the principal forces and dynamics – and their interactions – within a process is an essential prerequisite for any forward-oriented activities, such as *forecasting* or *predicting* future population or traffic growth or travel demand, policy formulation in order to set principles and guidelines for expected or desired future outcomes, and appropriate *interventions* or *planning actions* designed to achieve the desired policies. Analysis and prediction can take different forms. In transport planning and engineering, the most common approach is through increasingly sophisticated quantitative, multivariate modelling. Such exercises operate within given sets of assumptions and according to the algorithms and base data which form the inputs to the system. However, this book will be concerned with a broader, more conceptual form of analysis, addressing the role and nature of transport systems and their interactions with development processes in countries of the South. This should not be taken to mean that this approach is irrelevant to the North; far from it. Indeed, politicians and transport planners there would do well to stand back from their implicit assumptions and examine their options on a broader canvas.

Structure of the book

The remainder of this volume is divided into six chapters. Chapter 2 outlines some of the principal trends in changing transport technology and use, and their implications, across the Third World over the last century or so. Recent statistics on the growth and use of different modes of transport in selected countries of the South are then presented and compared with data on the North. The underlying reasons for such differences are explained in order to inform the more detailed coverage of specific issues in subsequent chapters.

Chapter 3 offers a concise discussion of the principal theoretical approaches to development and underdevelopment, and the role of transport therein. The shortcomings of a narrow technicist approach are highlighted and the value of interdisciplinary perspectives is emphasised with reference to contributions made by analysts working in different subject areas. The gulf between theoretical analysis and practical project implementation will be covered, together with the often crucial role played by external development agencies, e.g. the World Bank and regional development banks, and private consultants.

In Chapter 4 the issues of absolute and relative (in)accessibility and the impacts of transport provision are explored. While often portrayed as the forerunner of development, infrastructural expansion may serve the interests of the state in 'capturing the peasantry', may undermine the viability of hitherto relatively isolated economies, and/or serve as conduits for new diseases.

Chapter 5 outlines the diverse forms of urban transport, including those loosely classified as paratransit. Their growth or decline is related to economic development, changing average income levels and often inappropriate, fragmented and contradictory policies. Differences between modest or secondary cities and rapidly growing primate metropolises are then examined, together with issues of intra-urban variation and inequality. Discussion then focuses on the congestion, pollution, land use and social equity challenges which the current situation presents, as well as some relevant policy options and the respective constraints.

In Chapter 6 attention shifts to long-distance freight and passenger transport, where technological change has had a dramatic impact. Differentiation and specialisation are traced through road, rail, maritime and air transport, stressing some of the implications of international standardisation and stiffening competition for different categories of Third World countries and transport operators.

The final chapter draws together some of the central threads running through the book and examines the implications of the foregoing analysis for the foreseeable future against a background of increasing heterogeneity and divergent development prospects within and among countries of the South. In accordance with the conventions of this series of introductory texts, the book ends with a section of review questions for each chapter and suggestions for further reading.

2
Third World transport: the changing situation

Introduction

The impact of technological change in the transport sector since the late nineteenth century has been as dramatic as in other spheres of life. Four particular innovations stand out in this respect: the spread of railways, the popularisation of motor vehicles driven by internal combustion engines, the introduction and dramatic growth in longhaul air travel (especially by jet aircraft) and the unitisation (or containerisation) of sea freight. However, their adoption has not been universal and their impact has been far from uniform or neutral. While certain groups of people, whether defined socially, economically or geographically, have invariably gained from each of them, others have suffered or had their livelihoods and development prospects undermined.

The importance of this point cannot be emphasised strongly enough. Technology has generally been – and is still widely – perceived as being neutral in itself, and technical innovation therefore as an unqualified benefit and symbol of development. This view is deeply embedded within the so-called modernisation theory of development, arguably the politically most powerful and dominant paradigm in development practice since the Second World War. It has remained the virtually unchallenged orthodoxy within Transport Studies right through to the present, providing a straightforward legitimising rationale for those believing that the provision of transport is both a necessary and sufficient precondition for development. This will be discussed in greater detail in Chapter 3.

However, technology neither exists nor is utilised in a vacuum. Its development, introduction and use occur in specific social and economic conditions in particular localities. Moreover, it is now generally acknowledged that, whilst transport provision might be necessary, it is certainly not a sufficient precondition for development to occur. How one evaluates the impact of a transport innovation therefore depends on one's position relative to that innovation within a society; in aggregate terms such an exercise would require some form of social cost/benefit analysis or a planning balance sheet for the society or country as a whole. Importantly too, an *ex post* (i.e. retrospective) evaluation would frequently yield a very different verdict from the appraisal undertaken before implementation, because the range and relative magnitude or importance of the various impacts actually experienced may not have been accurately or comprehensively predicted.

An additional issue of central importance to Development Studies and hence the subject of this book is that the major modern transport technological innovations now found worldwide were developed in the North and exported very quickly to countries of the South, where social, economic, political and environmental conditions were or are usually very different. Scant regard has generally been paid by those responsible (in either the exporting or importing countries) to questions of appropriateness, affordability or maintainability of new transport technologies in these very different contexts, let alone the wider socio-economic impacts. This rather cavalier attitude is still common today and is reminiscent of the paternalism or worse that characterised the colonial era. For example, a penetrating study of the fashion for mass rapid transits in Southeast Asian metropolises examined the extent to which the phenomenon represents a form of 'technological imperialism' (Dick and Rimmer 1986).

On the other hand, some outspoken governments in the South respond angrily to such charges, claiming that if a new technology is good for the North it is good for the South. In other words, to prevent rapid or widespread adoption represents a form of discrimination and indeed imperialism, and would perpetuate the North's technological supremacy and hence international inequalities. Such arguments can quickly become emotive. The crucial issues are whether choice exists, who exercises it and on what terms, and whether all the relevant direct and indirect costs and benefits have been duly considered in that process.

In some respects the controversy over transport technology is comparable to that over tropical rainforest conservation, where Northern demands for the South to change its practices are challenged by some Third World leaders, perhaps most notably the Malaysian Prime Minister, Mahatir Mohamed, on grounds of double standards. They argue that the North destroyed a high proportion of its forests in the course of development over several centuries, but is now seeking to prevent the South from utilising its own resources for the same ends. Of course, many other considerations need to be taken into account, for example our greater state of knowledge and understanding of environmental processes and interrelationships; the global scale of many such processes and problems; the pressures facing indigenous communities and associated land rights issues; and the unprecedented level of globalised production processes and other economic, political and social interactions.

The widespread introduction of railways during the second half of the nineteenth century enabled far more rapid expansion of the European settler frontier and the 'taming' of the environment and subjugation of indigenous populations in existing or former colonies like the USA, Argentina, Mexico, Australia, Canada and India. This technological innovation also coincided with the late wave of new European colonisation in Africa, the so-called 'Scramble for Africa', with the result that even remote mineral deposits or areas deemed suitable for commercial plantations or arable cultivation were quickly linked by rail to coastal entrepôts and thence by sea to the imperial heartlands of Europe. The speed of such transport and the markedly greater volume and mass of freight able to be carried over long distances transformed the nature of such ventures and greatly intensified the impact of colonial rule. Pre-existing indigenous or colonial transport modes were quickly superseded, while new relations of production and mobility patterns were established. Territory, communities and their development potential were also increasingly differentiated in terms of their proximity to or remoteness from this new infrastructure (see Case studies A and B).

In certain more recent cases, of course, the increasingly closely integrated nature of the world system and world economy also left Third World countries little effective choice regarding the adoption of new technologies originating in the North if they wished to retain the active trading links on which they depend. The rapid shift to containerisation of non-bulk sea freight from the late 1960s onwards is an excellent example. Ports around the world had to invest in and accommodate the new technology within a very short period, making palletised general cargo ships, wharves and warehouses, as well as most of the stevedores employed to operate them, redundant for all except small local traffics and certain feeder services.

Furthermore, with increasingly sophisticated technologies and the associated safety issues, there may sometimes be little or no flexibility in terms of their introduction and use. Thus, for example, the worldwide regulatory authority in civil aviation, the International Civil Aviation Organisation (ICAO), sets strict safety and navigational equipment standards for recognised airports and aerodromes. Accession to the relevant treaties and compliance with their requirements is essential for recognition and the right to participate in the system. In other words, one cannot opt into some aspects of the system and technology but ignore others. A country is either in or out, regardless of the cost implications or perceived relevance of particular provisions and facilities to local conditions. In view of the importance of civil aviation to the international movement of freight and passengers – diplomats, bureaucrats, 'development' workers, businesspeople and tourists alike – not one country has opted to remain outside the scope of these treaties and the respective technological requirements.

Recent trends in transport use

Let us now examine recent worldwide trends in the provision and use of different modes of transport. Inevitably, the choice of variables and countries for comparison is somewhat restricted by data availability and in some cases by questionable accuracy, but the picture to emerge is reasonably consistent. In general terms, fixed transport infrastructure, such as railways, roads, airports and ports, is more extensive, of higher quality and better maintained in countries of the North than in the South. However, rates of growth in car and other motorised vehicle ownership, in the paved road network and in the tonnages of freight handled and moved, have been most rapid in upper middle-income countries, particularly the NICs (newly industrialising countries) and potential NICs. In other words, as ownership and use trends in some of these countries approach those of the North, differentiation within the Third World as a whole is increasing markedly.

Physical transport infrastructure

Rail networks

Notwithstanding the foregoing argument, growth in the absolute stock of transport infrastructure provided is by no means inevitable or irreversible. For example, railway networks have been contracting over the last thirty years or so in many (but by no means all) high-income countries of the North (Table 2.1), as a reflection of an ongoing intermodal shift in both passenger and freight traffic from rail to road. It is equally noticeable that Ireland, Germany, Spain and Japan, for example, increased their length of track significantly. Since the early 1980s, concern with liberalisation and privatisation of state owned and operated infrastructure under structural adjustment and economic recovery programmes (see Chapter 7) has seen underutilised branch lines and even some main lines closed in a few countries of the South. This is evident from Table 2.1 in respect of Kenya, Zimbabwe, Colombia (since 1970), Argentina (until 1980), South Korea (since 1970) and especially Brazil. Conspicuous in this group are several NICs and potential NICs, which suggests that rail transport and industrialisation may be inversely rather than directly related. However, inconsistent trends, with periods of both growth and decline, have been rather more common. Uganda, Egypt, Tunisia, Algeria, Costa Rica, Chile, South Africa and Uruguay illustrate this well. Several other countries, e.g. Ethiopia and Angola, experienced a decline in track length during the 1980s; it is unclear, though, to what extent this was planned as opposed to representing the effect of destruction or closure during the bitter wars raging there over that period.

However, Table 2.1 also reveals that the dominant trend among low- and middle-income Third World countries is still towards the laying of additional track. India, which already has the world's most intensively used railway system, continues to add to its network, with substantial growth recorded both there and in neighbouring Pakistan during the 1980s. Argentina and South Africa also added substantial lengths of track during the 1980s but the most dramatic increase took place in the Dominican Republic (not listed), where track length trebled, from 590 to 1,655 km. In some countries, like South Africa, sudden expansion can be explained by the exploitation of new mineral resources and/or the development of new harbours. Zambia has also recorded modest track increases despite the economic crisis prevailing since the mid-1970s and recent efforts to rationalise and commercialise Zambian Railways.

Table 2.1 Physical infrastructure (selected countries)

Country	Paved roads (km)				Railway tracks (km)			
	1960	1970	1980	1990	1960	1970	1980	1990
Low-income economies								
2 Ethiopia	–	1,935	11,320	13,198	1,090	1,090	987	781
6 Uganda	1,200	2,218	3,871	2,416	1,300	5,895	1,145	1,241
9 Malawi	485	750	1,905	2,320	509	566	782	782
10 Bangladesh	–	3,610	4,283	6,617	–	–	–	2,892
18 India	254,446	324,758	623,998	759,764	56,962	59,997	61,240	75,333
19 Kenya	–	2,570	5,558	6,901	6,558	6,933	4,531	2,652
20 Mali	–	1,596	1,795	5,959	645	646	641	642
21 Nigeria	–	15,216	30,021	31,002	2,864	3,504	3,523	3,557
26 Pakistan	16,860	24,776	38,035	86,839	8,574	8,564	8,815	12,624
33 Zimbabwe	–	8,474	11,788	12,896	3,100	3,239	3,415	2,745
36 Eygpt, Arab Republic of	–	10,059	12,658	14,601	4,419	4,234	4,667	5,110
37 Indonesia	10,973	21,073	56,500	116,460	6,640	6,640	6,637	6,964
40 Sudan	–	332	2,975	3,419	4,232	4,756	4,787	4,784
42 Zambia	–	2,877	5,576	6,198	1,158	1,044	1,609	1,894
Middle-income economies								
Lower middle-income								
43 Côte d'Ivoire	829	1,258	3,057	4,216	624	656	680	650
48 Senegal	–	2,097	3,445	4,000	977	1,186	1,034	1,180
56 Congo	–	378	561	985	515	802	795	510
67 Colombia	2,998	5,980	11,980	10,329	3,161	3,436	3,403	3,239
72 Tunisia	6,845	9,106	12,278	17,509	2,014	1,523	2,013	2,270
74 Algeria	–	32,963	38,929	44,191	4,075	3,933	3,907	4,653
75 Thailand	2,740	9,656	23,613	39,910	2,100	2,160	3,735	3,940
79 Costa Rica	–	1,400	2,424	5,600	665	622	865	696
85 Chile	2,604	7,411	9,823	10,983	8,415	8,281	6,302	7,998
Upper middle-income								
89 South Africa	–	33,115	46,634	51,469	20,553	21,391	20,499	23,507
92 Brazil	12,703	50,568	87,045	161,503	38,287	31,847	28,671	22,123
98 Uruguay	1,473	6,002	9,792	–	3,004	2,975	3,005	3,002
102 Argentina	22,712	33,375	52,194	57,280	43,905	39,905	34,077	35,754
106 Korea, Republic of	733	3,618	15,587	34,248	2,976	3,193	3,135	3,091
High-income economies								
116 Australia	80,800	167,920	244,086	263,527	42,376	43,380	39,463	40,478
117 United Kingdom	319,314	334,132	339,804	356,517	29,562	18,969	18,028	16,629
118 Italy	–	262,188	285,319	303,906	21,277	20,212	16,133	25,858
119 Netherlands	70,000	78,551	92,525	92,039	3,253	3,148	2,880	3,138
124 France	626,400	690,950	730,697	741,152	39,000	36,532	34,382	34,593

Table 2.1 continued

125 Austria	32,063	94,832	106,303	125,000	6,596	6,506	6,482	6,875
129 Denmark	41,283	50,676	68,909	71,063	4,301	2,890	2,461	3,272
131 Japan	37,785	152,033	511,044	782,041	27,902	27,104	22,235	23,962
Selected economies not included in main World Development Indicator tables								
Angola	–	5,351	–	7,914	3,110	3,043	2,952	2,523
Iraq	7,316	4,773	14,166	26,040	2,019	2,528	1,589	2,372
Liberia	–	322	1,800	2,279	493	450	493	493
Suriname	459	–	–	2,379	136	86	167	166
Swaziland	–	182	447	688	225	220	295	316
Zaire	–	2,110	2,175	2,800	5,074	5,024	4,508	5,088

Source: World Bank (1994: 140–5)

Paved roads

The trend on this variable is unambiguously towards increasing network length for all income categories of country (Table 2.1). Some very dramatic relative increases have been recorded over the course of a single decade, for example in Burundi (not listed), Pakistan, Indonesia and Costa Rica from 1980 to 1990. The 1990 figure for Burundi (1,011 km) was treble that for 1980 (365 km), while in the other three countries the paved networks more than doubled. It is conspicuous that many of the poorest countries experienced the highest relative increases. This reflects the low base from which they started and the poor quality of existing roads. However, by no means all of this gain represents an increase in the total road network. A high proportion is accounted for by the upgrading and tarring of existing gravel roads, often funded by external donors as part of official development assistance programmes targeting high-volume routes. On the other hand, many integrated rural development programmes in low-income countries have included feeder road components (see Chapter 4). In absolute kilometre terms, India achieved the most remarkable paved road increase, from 623,998 to 759,764 km during the 1980s. It is worth noting that, in contrast to the decline in their rail networks, both Mozambique (not listed) and Ethiopia, the two poorest countries in terms of GNP per capita, were able to extend their tarred road systems significantly despite their protracted respective wars. There are a handful of exceptions, perhaps most conspicuously Uganda and the Philippines, where the total paved road network apparently declined substantially over the decade. If these figures are accurate, they reflect the degrading of previously tarred roads, through lack of maintenance, to the point where they have been classified downwards to gravel status.

Relative indicators of infrastructural provision

While absolute physical indicators of the type discussed above provide some interesting information, they are of only limited and descriptive value. They tell us nothing about the interregional or urban versus rural distribution of the networks, nor how dense they are and what proportion of the population has access to them. It would therefore be far more helpful, for example, to relate these to population, income level or geographical data to yield network densities, or the availability of transport services. This has been done in Table 2.2, which provides road density in km per million people; the percentage of paved roads in good condition; rail traffic in km per million $ of GDP; and the percentage of the diesel locomotive stock actually in use on the railways. Of course, all these figures are national averages which, whatever their degree of reliability, conceal often substantial interregional and/or urban–rural differences. Being relative variables, there are two potential sources of error in each, e.g. with the figures for both road network length and total population in the first variable.

Access to roads

The road density variable represents a simple attempt to indicate average potential access to a road, regardless of its condition or hierarchical status, by the population of a country. Other useful indicators of proximity or access, e.g. the percentage of population within 1 km of a road, are seldom available. The second road variable is an indicator of tarred road quality. 'Good condition' is defined in the *World Development Report 1994* as roads substantially free of defects and requiring only routine maintenance. This is a function of the newness and use made of roads as well as the level of maintenance of existing roads in the network. The percentages relate to the total tarred road networks in the respective countries as given in Table 2.1, in other words, principally trunk and major urban roads. As explained above, in most countries of the South, tarred roads represent only a comparatively small proportion of the total road network and may be less directly relevant to the majority of rural inhabitants than the condition of their nearest rural

Table 2.2 Relative measures of infrastructural provision (selected countries)

		Paved roads		Railways	
		Road density (km per million persons) 1988	Roads in good condition (% of paved roads) 1988	Rail traffic (km per million $ GDP) 1990	Diesels in use (% of diesel inventory) 1990
Low-income economies					
2	Ethiopia	84	48	–	–
6	Uganda	118	10	–	49
9	Malawi	278	56	43	77
10	Bangladesh	59	15	41	73
18	India	893	20	593	90
19	Kenya	278	32	120	52
20	Mali	308	63	106	44
21	Nigeria	376	67	17	20
26	Pakistan	229	18	168	79
33	Zimbabwe	1,389	27	505	54
36	Eygpt, Arab Republic of	302	39	394	93
37	Indonesia	160	30	–	74
40	Sudan	98	27	27	29
42	Zambia	751	40	294	44
Middle-income economies					
Lower middle-income					
43	Côte d'Ivoire	357	75	35	58
48	Senegal	542	28	78	62
56	Congo	584	50	170	56
57	Morocco	618	20	141	88
67	Colombia	309	42	5	35
72	Tunisia	1,177	55	123	50
74	Algeria	1,366	40	85	99
75	Thailand	513	50	76	72
79	Costa Rica	1,059	22	–	–
85	Chile	753	42	48	57
Upper middle-income					
89	South Africa	–	–	987	88
92	Brazil	704	30	60	62
98	Uruguay	2,106	26	15	56
102	Argentina	858	35	161	49
106	Korea, Republic of	236	70	–	89
High-income economies					
116	Australia	25,695	a	62	–
117	United Kingdom	6,174	a	66	–
118	Italy	5,254	a	90	80
119	Netherlands	6,875	a	73	83

Table 2.2 continued

	Road density (km per million persons) 1988	Roads in good condition (% of paved roads) 1988	Rail traffic (km per million $ GDP) 1990	Diesels in use (% of diesel inventory) 1990
124 France	14,406	a	146	93
125 Austria	14,101	a	209	90
129 Denmark	13,775	a	93	–
131 Japan	6,007	a	144	87

Note: a: 85% or more are in good condition
Source: World Bank (1994: 224–5)

access road. Unfortunately, such data are seldom available, not least because of the wider range in gravel road conditions.

Although now somewhat dated, the 1988 road data in Table 2.2 are the most recent available. They reveal three noteworthy patterns:

1   There is wide variation on both variables within each income category of country. Road densities in the low-income countries listed range from a mere 59 km per million people in Bangladesh, a figure higher than only Chad and Mali (56 and 21 km/mn respectively) to 1,389 km/mn in Zimbabwe. In the lower middle-income category, Colombia (309 km/mn) is the lowest and Algeria (1,366 km/mn) the highest, although Jamaica (not listed) achieved 1,881 km/mn. Among upper middle-income countries, the range is from a surprisingly low 236 km/mn in the Republic of Korea to 2,106 in Uruguay and an impressive 10,269 in Venezuela (not listed). This is a very atypical figure, exceeding that for many high-income countries, including Japan. Among high-income countries, the difference between Japan and Australia reflects the physical size and average population density of the respective countries, two variables which outweigh the difference in total populations between them. Similar variations are evident in the percentage of tarred roads in good condition: from a mere 10 in Uganda to 67 in Nigeria among low-income countries; from only 7 in Guatemala (not listed) and 28 in Senegal to 75 in Côte d'Ivoire among lower middle-income countries; and from 26 in Uruguay to 70 in the Republic of Korea (and 96 in Mauritius – unlisted) in the upper middle-income category. Only within the high-income category do all countries listed have a figure of over 85 per cent.

2   There is no clear relationship between the two variables for individual low- and middle-income countries. For example, although Bangladesh scores very low on both, Zimbabwe provides a sharp contrast, with
the highest value for road density but one of the lower percentages for
tarred road quality among low-income countries. Algeria comes out well
on the first variable in the lower middle-income category but only
average on the second variable. Among upper middle-income countries, South Korea has a low road density but a high proportion of
tarred roads in good condition.

3   The values within each category of low- and middle-income country
overlap with those in other categories. With the exception of a handful
of cases, high-income countries have higher values than upper middle-
income countries on both variables.

Rail utilisation and availability

The third and fourth variables in Table 2.2 are proxy indicators for the
importance of railways to economic production in the formal sector (rail
km per million $ of GNP) and the provision and maintenance of diesel
locomotives (percentage of diesel locomotives in use). The rail traffic
units used in the third variable are the sum of passenger-kilometres and
freight tonne-kilometres. These datasets are very incomplete, making
generalisation difficult. It is important to note, however, that part of the
reason for the data deficiency lies in the choice of variable. Many
countries use electric and/or steam locomotives instead of or in addition
to diesel ones. It is impossible to tell from the data provided in the *World
Development Report 1994* whether the gaps in the table reflect the
absence of data or lack of use of diesel locomotives. Even where figures
are given for that variable, we are unable to deduce whether diesels
represent all or only part of the railway locomotive stock. It would
therefore have been far more sensible and useful to use the percentage
of *all* railway locomotives. The data as they stand are, frankly, of little
value. This is particularly unfortunate because rail transport, both intra-
urban and interurban, forms an integral part of many poor countries'
transport systems and is utilised particularly by the poorer segments of
the population, for example, long-distance migrants and commuting
workers.

The foregoing trends highlight the diversity of conditions among low-
and middle-income countries and the complexity of development issues.

It is very difficult to generalise, because so much depends on the particular variable or variables being examined, as well as the completeness and accuracy of the data. The road data certainly indicate that the ranking of countries according to GNP per capita provides a poor indication of infrastructural provision and availability, let alone other social and economic dimensions of development.

Non-motorised transport: bicycle usage

The role of bicycles, one of the most common forms of non-motorised vehicle, differs widely over time and between societies. In the North, they were (and in some places still are) used primarily as a means of transport to work and for shopping by working-class people. As average incomes and car ownership increased, bicycles became associated more with leisure and recreational use, except in small and particularly university towns and in flat regions like the Netherlands, where they remain popular with a wide spectrum of the population. In recent years, however, they have been used increasingly for commuting and to provide courier services in major cities, offering a pollution-free, cheap and efficient means of transport. New designs like the mountain bicycle have emerged for specialised leisure use.

In Third World societies, use of and attitudes to bicycles differ widely, in accordance with economic conditions, the terrain and cultural values. Data are patchy, but Figure 2.1 provides a broad indication of continental regional differences in cycle ownership in the early to mid-1980s. Ownership was highest in China (although car ownership there has been rising significantly over the last decade) and Asia in general, with Latin America a long way behind. Given that Africa is now the world's poorest continental region, the extremely low figure is rather surprising. Clearly, there is no straightforward or linear relationship between per capita income and bicycle ownership. Once again, therefore, even basic statistics reveal the limitations of a narrow and economistic modernisation approach. In terms of this, cycle ownership would be predicted to rise with income up to a threshold at which bicycles become regarded as 'inferior goods', and then taper off. This profile probably does not even hold within individual countries. Explanation must be sought elsewhere.

The importance of cultural attitudes and local traditions in this regard is illustrated by recent research in Ghana, showing that 34.2 per cent of the estimated 200,000 bicyles countrywide in the mid-1980s were to be

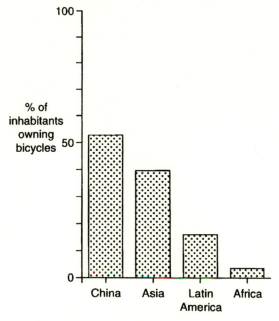

Figure 2.1 Bicycle ownership by Third World region
Source: Based on data in Grieco, Turner and Kwakye (1994) and Turner, J. (1994)
Transport Patterns and Travel Behaviour in Urban Ghana – Pilot Stage, Project Report
PR/OSC/047/94, Crowthorne: TRL

found in the Northern Region, 43.8 per cent in the two upper regions and
a mere 0.7 per cent within greater Accra, the capital city. Within Accra
too, and even within low-income areas of the city, ownership and use are
very uneven. In the area with a particularly high concentration of
Northern migrants, ownership is estimated at 62 per cent of households; in other low-income areas it is low and cycling is mainly local.
This variation reflects a Northern tradition of cycling (Grieco *et al.*
1994: 1), coupled with perceptions elsewhere, probably derived in part
from colonial attitudes, that cycling is an inferior or undignified mode of
transport. Under the circumstances, even if road safety were to be
improved for cyclists, no significant increase in use of this mode would
be anticipated. Attitudinal change would be a prerequisite; merely
promoting cycling as economically and environmentally sustainable in
a manner comparable to the common current practice in parts of the
North, would not have a noticeable impact.

Unfortunately, comparative statistics on non-motorised transport like bicycles are particularly fragmentary, precluding any firm conclusions regarding their extent and relative importance in different countries or regions within them. Somewhat impressionistically, therefore, and perhaps surprisingly, it seems that in countries where bicycle usage is low for reasons other than topography, bicycles tend to be no more common (and often less so) in intermediate and smaller towns, to which they are particularly appropriate in view of less problematic levels of motorised traffic and shorter travel distances, than in large metropolises. Levels of use may actually be higher in rural areas because of the inadequacy of appropriate public transport.

Of course, it is also true that the purchase of a bicycle, especially if new, represents a substantial capital investment which may be beyond the means of many poor households. For those who can afford it, opportunities may exist to recoup some or all of the outlay by hiring out the bike when it is not in use. In rural areas, ownership tends to be higher among those with access to wage incomes, especially migrant workers, certain categories of public sector employees and successful small farmers.

Hi-tech travel trends: civil aviation

The level of civil aviation traffic, of both passengers and freight, is rather more closely related to average living standards than most variables, but is also subject to significant economic cyclical fluctuations. It is not therefore surprising that Africa accounted for only 2.2 per cent of global passenger numbers on scheduled flights of airlines registered in the respective countries, 2.2 per cent of passenger-kilometres travelled, and 2.1 per cent of total freight tonne-kilometres in 1992 (Table 2.3), despite being home to over 12.2 per cent of the world's population in 1992. The figures for Latin America and the Caribbean (i.e. South America plus all of North America except the USA and Canada) are somewhat more in line with the region's share of global population, namely 4 per cent of global passengers, 4.8 per cent of passenger-kilometres and 4.8 per cent of global tonne-kilometres as against 8.3 per cent of the world's population. However, the markedly higher traffic volume in Asia reflects both total population and, crucially, the rapid recent economic growth rates recorded in Pacific Asia. The importance of the USA and Canada (5.2 per cent of the world's population but 42.7 per cent of total passengers carried, 42.4 per cent of global passenger-kilometres and 39.0 per cent of total tonne-kilometres

Table 2.3 Scheduled civil air traffic

	Total			International		
	1980	*1990*	*1992*	*1980*	*1990*	*1992*
World						
Passengers carried[1]	645,201	1,027,320	1,123,620	160,698	275,897	298,986
Passenger-km[2]	929,004	1,652,297	1,899,054	457,512	875,625	981,439
Total tonne-km[3]	113,486	210,379	238,812	63,140	128,729	142,985
Africa						
Passengers carried[1]	21,235	26,694	25,361	9,003	12,652	12,636
Passenger-km[2]	29,724	42,337	42,564	22,434	33,465	34,629
Total tonne-km[3]	3,544	4,957	5,013	2,811	4,069	4,229
America, North[4]						
Passengers carried[1]	336,904	508,282	507,117	38,700	66,544	66,035
Passenger-km[2]	467,244	812,497	837,185	112,724	241,337	259,637
Total tonne-km[3]	53,926	93,276	96,300	14,691	31,573	33,811
America, South[5]						
Passengers carried[1]	34,008	41,457	43,668	6,049	7,547	12,194
Passenger-km[2]	38,628	56,853	59,832	19,547	31,109	36,378
Total tonne-km[3]	4,790	7,660	8,304	2,907	4,932	5,815
Asia						
Passengers carried[1]	106,119	209,139	235,885	34,301	67,643	78,036
Passenger-km[2]	158,192	344,494	402,512	108,019	241,758	281,115
Total tonne-km[3]	20,458	48,355	55,803	15,977	38,735	44,684
Europe						
Passengers carried[1]	128,802	216,464	217,498	69,391	115,079	119,741
Passenger-km[2]	203,021	342,866	369,410	173,908	290,285	314,262
Total tonne-km[3]	27,042	49,535	52,677	24,200	44,381	47,244
Oceania						
Passengers carried[1]	18,133	25,286	37,770	3,254	6,432	7,361
Passenger-km[2]	32,195	53,250	71,310	20,881	37,672	43,053
Total tonne-km[3]	3,727	6,596	8,622	2,554	5,039	5,770

Notes: [1] Passengers in thousands   [4] Includes Central America and Carribbean islands
[2] In millions   [5] Mainland only
[3] Excludes mail and excess baggage
Source: UN (1994) *Statistical Yearbook 1992*, New York, 675–90

of freight) and Europe is long established, although suffering the effects of recession during the late 1980s and early 1990s. In relative terms, however, their dominance is declining as Asian traffic increases.

Table 2.3 reveals clearly that the pace and significance of growth in Asia have been by far the most rapid in the world, reflecting the region's remarkable economic performance: passenger numbers virtually doubled, and passenger-kilometres and tonne-kilometres more than doubled in the decade 1980–90, trends that were continuing into the early 1990s. By this time, Oceania was also experiencing the effects of this surge, with 1990–2 growth rates on all three variables actually outstripping those in Asia. By contrast, 1992 data on Africa were only marginally higher than for 1990; indeed, passenger numbers were lower, although the 1992 figure represented a recovery from a marked decline in 1991.

Of course, such aggregate continental data conceal major differentiation between and within subregions and countries. In Africa, for example, airlines registered in South Africa carried considerably more passengers in 1992 than those of any other country, accounting for 18.5 per cent of the continental total. Egypt was in second place, with 14.2 per cent, followed by Algeria (14 per cent), Morocco (8.6 per cent), Libya (5.3 per cent) and Tunisia (4.9 per cent). These six, out of the fifty countries included in the UN dataset, therefore accounted for 65.5 per cent, i.e. virtually two-thirds, of the continental total. At the same time, with the exception of South Africa, the other five together constitute the Maghreb (or North Africa, although the UN also include Sudan within their definition of this region), which therefore accounted for 47 per cent of the total for Africa.

Although, with the exception of Egypt, the six countries cited were classified by the World Bank as middle-income countries in 1992 (upper middle-income in the cases of Libya and South Africa), this alone does not explain the importance of their airlines and air travel. The figures reflect some combination of several factors: the existence of significant local middle classes; substantial foreign commercial activity; large numbers of nationals working abroad as migrants or undertaking business travel; the size, reputation, cost, and hence relative market shares, of locally registered airlines; and/or the existence of important foreign tourist industries. Although subdivision of the data for these countries into domestic and international passengers reveals significant differences, the proportion of international passengers is substantial in each case.

Finally, it should be pointed out that data such as these can vary quite markedly depending on the source. This may arise from different calculation procedures (rounding and aggregation errors and the like), varying degrees of accuracy, or different definitions. The last-mentioned is particularly important but often overlooked. For example, the air passenger data used here refer to scheduled flights of airlines registered in the respective countries. The figures thus obtained are very different

from the data published by ICAO, which list passengers by airport
(*Airport Traffic*, Series AT) or sector (*Flight Stage*, Series TF), in other
words passengers passing through each airport and using all direct routes
into each airport respectively. As it happens, Johannesburg and Cairo
airports do also have the highest passenger throughputs in Africa, but it
would have been very misleading to imply or infer this from the figures
just presented. It is therefore important to specify the nature of data used
and to exercise appropriate care in the interpretation of conclusions
derived from them.

Of course, in developmental terms, air travel is totally irrelevant to the
great majority of the population in most Third World countries. It
provides fast, high-cost, high- (and generally imported) technology
connections between large cities. This is affordable only by the wealthy
elites and middle classes, who are also almost exclusively the ones who
make up the cohort of business travellers, and is generally used only for
high-value freight. Moreover, it has dramatically changed the time
geography of travel, providing faster, more comfortable and usually
more reliable connections between distant localities, even across
national and continental boundaries, than often exist between places
within the same country or even subnational region which are far closer
together in terms of physical distance. This has greatly facilitated the
growth and integration of transnational corporations, intergovernmental
organisations, international agencies and other organisations. Together
with telecommunications, air travel is vital to the global 'information
economy'. These trends have led to changing geographies of production,
distribution and consumption, different delineations of the world with
differing degrees of integration with, or marginality to, such technologies and processes. As indicated in Chapter 1, the Third World is
becoming increasingly differentiated in these terms, with Singapore,
Hong Kong, Taiwan, the Republic of Korea and a few others at one
extreme and countries like Mali, Chad and Haiti at the other. Some of
these ideas will be returned to in later chapters.

In wider terms, airports and the airline industry themselves also
provide substantial employment for different skill categories while
contributing to the growth of tourist and other industries. Even the
most menial airport jobs are highly sought after, and access to them
may be greatly influenced by personal or political connections. Airports
are also frequently located adjacent to low-income areas in the urban
periphery; airport perimeters may also offer relatively attractive sites for
the erection of shanties or other squatter dwellings because of proximity

to employment and/or a reduced risk of demolition by the authorities or property developers. In Bombay, for example, the airport perimeter fence, which passes close to the end of the runway, marks the boundary of a large shantytown. Under such circumstances, it is the poor who bear a substantial burden in terms of noise and aerofuel pollution and the highest risk in the event of an accident.

Conclusions

It is very easy to see technology in rather simplistic terms as representing something undoubtedly useful and beneficial. Fascination with gleaming steel or the sense of power in a new type of engine or locomotive, coupled with the convincing advertising of corporate sales representatives, can obscure important issues about how that technology is used, and the effects which that usage might have. As this chapter has shown, the impacts are often complex and some may be unanticipated. While some people, and the interests they represent, undoubtedly do benefit, others may actually lose their livelihoods. The precise developmental impacts of technological change therefore depend on the balance of gains and losses, both directly within the transport industry itself and indirectly in the wider social economy.

In sketching the context for the rest of the book, this chapter has examined a selection of recent data on transport and travel in the South, and compared and contrasted this with trends in the North. The differences are generally still substantial, although the clear-cut divisions between groups of countries are becoming increasingly blurred. There is considerable diversity within groups of countries too, whether these are defined on the basis of per capita GDP or geographical region. The evidence revealed that there is no direct or linear relationship between the provision of transport infrastructure, or the use of most forms and modes of transport, and average national income. Numerous other factors, geographical, social, political and economic, are as important. The case of bicycling was used to illustrate the centrality of cultural values and perceptions to understanding differences between apparently similar areas in Ghana. Such attitudes are often far more powerful influences on behaviour than economic cost and notions of sustainability.

The one example discussed which does relate closely to income is air travel, which involves more sophisticated technology and remains beyond the reach of most Third World citizens, despite the falling real cost in recent years and growth of mass air tourism from countries of the

North to 'exotic' locations in the tropics and subtropics. Mass tourism itself is a mixed blessing, being promoted vigorously by the tourist industry as well as by many Third World governments as a leading element of their development strategies, but having a high import content in poor countries and spawning a number of dubious practices and undesirable social effects (Lea 1988).

There are also great variations in the extent and accuracy of data coverage in different regions and countries and for different transport modes. Statistical reporting is, to a significant extent, a function of national income, as well as prevailing conditions. Generally, however, the routine collection of data is biased towards officially registered, motorised and 'modern' forms of transport, while paratransit, cycling and walking, for example, are usually neglected. Unless specific studies have been undertaken, it is impossible even to estimate the number or importance of such forms in individual cities, let alone entire countries. Although some attitudes are gradually changing or at least softening, non-motorised forms of transport are still widely perceived by governments and national elites to symbolise past traditions, a lack of modernity and a source of traffic congestion which should be phased out of large urban areas. Nevertheless, they remain important to very many urban residents and new migrants.

Key ideas

1 The adoption of new transport technologies generally has differential economic and social impacts in any local context.
2 Transport planners and policy makers seldom take account of such issues or of the important differences in prevailing conditions between the North, where the technologies are generally developed, and those in the South when the technologies are imported there.
3 There are substantial differences in infrastructural provision and use between countries, even within the same subcontinent or World Bank income category.
4 Absolute measures of infrastructural provision, e.g. length of rail or road networks, are of limited descriptive value; relative indicators are generally more useful as measures of average access to or potential use of transport.
5 Although many governments, transport professionals and national elites consider non-motorised transport to be inferior, old-fashioned, and a source of traffic congestion, such modes often form important

components of local transport systems and may be more appropriate to prevailing conditions than more sophisticated imported technologies.
6 Nevertheless, it is difficult to adopt certain newer technologies and their associated infrastructure gradually or in stages. Civil aviation is a good example of such 'all or nothing' situations.
7 Charges of 'technological imperialism' may be countered by claims that 'what's good for the North is good for the South'.

3
Transport and development: exploring the linkages

Introduction

In this chapter, we move beyond the description and discussion of transport trends and their development implications to explore the underlying causes and processes. This is undertaken with reference to alternative theoretical perspectives on the relationship between transport and development. Following a brief summary of the relevant theories, their application and influence within Transport Studies and Development Studies will be outlined, with reference to specific examples and case studies. The considerable gap between such theoretically informed work and the implicitly atheoretical approach adopted by many international consultants and organisations will then be explored in the context of the often great influence of and crucial role played by such agents and agencies.

The importance of theories is that they are simplifications or abstractions which highlight key variables or relationships in an effort to explain and understand the complexities of the real world. A true theory, moreover, offers more than a description of such relationships or processes. It should offer an explanation of how the current situation arose and be capable of predicting future changes or developments of the system. In other words, it must have explanatory and predictive power. Otherwise it is merely a descriptive relation. Put most simply, a theory represents a 'way of seeing', a view of the world, an aid to ordering and organising the world.

Such a perspective is a prerequisite for coherent and consistent intervention aimed at altering the existing situation and creating a more desirable future outcome. Otherwise, actions risk being uncoordinated and even contradictory. More important, even, is that a theoretical or conceptual view provides guidance on *how* to intervene and *what* actions are likely to prove appropriate. In other words, successful intervention requires an understanding of the present situation and how it has arisen, one or more desired future scenarios, and an ability to predict the likely impact of changes introduced to the underlying relationships in order to achieve the end result. This is the essence of policy formulation and planning.

Of course, we all have notions in our minds of our present situations, of 'how the world works', and our desires or aspirations for the future. However, not everybody has the experience, insight, training or motivation to work out how to set about achieving those aspirations. Our world views, beliefs and actions are often implicit rather than explicit, are formed by different blends of emotional and intellectual or conscious input, and are probably not entirely consistent. Professional planners and forecasters receive formal education and training designed to provide them with the necessary concepts, theories, methods and techniques to undertake their work dispassionately and consistently, in accordance with certain principles or guidelines. Part of this education should involve the requirement that all assumptions and value judgements as well as the terms of reference are made explicit at the outset. In practice, this often does not happen.

This situation may reflect the political environment in which the planners are working, where the outcome is a foregone conclusion, having been predetermined by political command, pressure or other process, perhaps involving corruption, undue patronage or clientelism. In other contexts, the data, information and resources available to the planners may be so inadequate as to make their task unrealistic, with the result that they take a short cut or deliver the least disruptive or cheapest option without due investigation. However, at least as important is the nature of education and training received by planners, both within the transport field and more generally.

In countries of the North, one particular theoretical framework or world view, namely modernisation theory, and the tradition of economic thought known as neoclassical economics from which it is derived, have long been so dominant in the social sciences that they are taken for granted. In the engineering and physical sciences, the dominant paradigm or theoretical perspective has long been that of 'rational science' or 'positivism', the belief that the key to understanding lies in empirical observation and measurement of material phenomena and processes. These two traditions are natural counterparts and have become mutually reinforcing. Their impact in a subject like Transport Studies, which is dominated by a combination of economics and engineering, has been particularly great. They also form the conceptual foundation of the major international financial institutions and development agencies, like the IMF, World Bank and regional development banks, and most UN agencies.

Through the establishment of colonial educational institutions and professional associations, and the ongoing training of many Third World professionals in countries of the North, these paradigms tend to dominate in many Southern countries as well, despite the often very different conditions prevailing there and sometimes radically different indigenous intellectual or scientific traditions. On the other hand, as elsewhere, some social scientists in countries of the South have been prominent critics of conventional 'Western' wisdom and have sought other paths to 'development'.

In recent years, various strands of social-scientific critique have emerged, challenging the whole notion that we should be concerned with finding any one single, overarching or 'grand' theory that is 'best'. This latter intellectual project is claimed to characterise modernism, the movement or paradigm centred on the search for and preoccupation with 'modernity'. Development, or more accurately developmentalism and much of Development Studies, has been criticised as being part of that project. Instead, postmodernism, postcolonialism and postdevelopmentalism, as these critiques are variously known, advocate less monolithic, more pluralistic, 'local' and even anarchic approaches, which reduce the dominant role of the state and its agents in setting agendas and planning. These fundamentally different perspectives have, naturally, proved very controversial, although as yet they have had a minimal impact in the transport sphere. Nevertheless, they will be summarised below and their potential evaluated.

The impact of modernisation theory

Modernisation theory has been the most widespread and powerful paradigm – in the sense of the influence of its adherents rather than of its innate explanatory power – in Transport Studies, and probably also

Development Studies, since the Second World War. Grounded in neoclassical economics, it had older roots in economics and sociology but became articulated as a coherent framework during the postwar era of reconstruction and development. Economic development was equated with development, a notion now widely discredited in favour of broader conceptions (see Chapter 1).

The object of development was deemed to be the pursuit of modernity through the modernisation of a country or region's economy through industrialisation along the lines of the Western experience. This could be achieved in any country by following a set of prescriptions derived from Euro-American experience, and designed to promote the expansion of large-scale, commercial economic development at the expense of 'traditional' or 'primitive', subsistence-oriented activities. Around the time of decolonisation in Asia and Africa, during the first two decades after the end of the Second World War, the colonial powers sought to establish some of the essential infrastructure and facilities that the new independent states would require, and which had been ignored or neglected during colonial rule. Until then, the colonies had been exploited very largely for the benefit of the metropolitan power rather than 'developed' with any inherent concern for the needs or priorities of the indigenous populations.

These new efforts were continued after independence in the form of aid, now formally known as official development assistance, from their former colonising countries and also multilateral agencies like the World Bank. These aid programmes were inspired by and modelled on the success of the US government's Marshall Plan for the reconstruction of postwar Europe, which had been ravaged by the years of fighting and aerial bombing. The Plan comprised the linked economic and political objectives of rebuilding Western Europe's shattered infrastructure and economies, and thereby seeking to ensure that Soviet-style Communism would be unattractive to its peoples. It also had the important motive for the USA of guaranteeing employment and exports for its own economy during the potentially difficult transition from a war footing back to civil production.

The results in Europe certainly were spectacular, as city centres, urban transport systems and industrial capacities were rebuilt using the most modern technologies and materials. It was then felt that if such large-scale, capital-intensive programmes had achieved their objectives so admirably in Europe, the same could surely be achieved in the South, where no war damage on a comparable scale had to be repaired. The 'problem' was rather perceived as one of initiating and then broadening 'modern' – and especially industrial – economic production.

The intention was to accelerate the newly independent countries through an industrial transition of rapid modernisation, the benefits of which would readily be evident in terms of improved living conditions and standards. While initially perhaps concentrated in one or more large urban centres for reasons of economic efficiency and to take advantage of agglomeration and localisation economies (in other words, cumulative causation), these would 'trickle down' or spread through the urban hierarchy and rural areas, progressively replacing traditional lifestyles, modes of production and the poverty with which they were (often entirely erroneously) associated.

Perhaps the best known exposition of modernisation theory is that popularised by Walt Rostow, who put forward a model of economic development comprising four successive stages, from pre-industrial to post-industrial, through which developing countries should pass. His conception was very limited, and the notion of a single linear path along which all countries that attained 'lift-off to self-sustaining growth' would pass has been rightly discredited as unrealistic. This is typical of the 'universalising' assumptions underpinning the theory as a whole, whereby Western experience and values are simply extrapolated to the rest of the world without regard to fundamental social, cultural or politico-economic differences. Actually, such differences were held to be irrelevant: the primary importance of economic development meant that it should be the immediate focus of intervention and change, which would in due course work through the societal fabric in any particular context, tranforming values and behaviour quite rapidly in what was seen as an unproblematic manner.

The evidence in favour of modernisation theory, e.g. the widespread diffusion of certain material symbols of modernity like transistor radios, denim jeans and Coca-Cola, is at best partial and patchy. There is also much literature documenting how diffusion and trickle-down have failed in the Third World, with little substantive change or improvement in standards and living conditions for a large proportion of the population, within the core region(s) as well as in the periphery. On the contrary, many previously self-sufficient or self-reliant communities have been reduced to impoverished dependence. Nevertheless, the power of the paradigm has persisted among transport and planning professionals, international agency staff and many Third World leaders and elites alike. This point warrants emphasis: the lure of greatly increased material wealth, encouraged by the demonstration effect of the North, proved highly attractive to most of the political leaders and elites in newly independent states. They enthusiastically embraced the theory and economic policies designed to implement it.

In part this enthusiasm reflects the simplicity and widespread appeal of the theory's overtly capitalist logic. To label this a case of Western imperialism is too simplistic and misleading, despite the undoubtedly unequal distribution of gains between North and South. Another crucial factor contributing to this enthusiasm was the global mood of economic optimism spawned by the postwar boom from the 1950s until the late 1960s, when rates of economic growth in the North and South alike were unprecedented. None of the alternative theoretical perspectives or subsequent economic upheavals, such as the impact of the recession of the late 1970s and early 1980s or the debt crisis, have dented the conviction of a substantial proportion of transport professionals and others. On the contrary, the demise of socialism in Europe and the Third World has reinforced the belief that such alternatives to growth through capitalist accumulation were fundamentally flawed.

Geographers, most notably John Friedmann, translated Rostow's model into spatial terms, developing a four-stage model in which a single, dynamic and modernising urban core expanded through the urban hierarchy and across the rural periphery, progressively reducing urban–rural disparities and producing a homogeneous, fully integrated and modern development space (Figure 3.1). Clearly, the role of infrastructure and reliable, efficient transport in ensuring the movement of people and goods is central, and is indicated in the model by the arrows showing the direction and nature of movement. In the early stages, centripetal forces and flows tend to drain the periphery, concentrating resources and skilled, able-bodied workers in the core; however, once diseconomies of scale begin to outweigh agglomeration economies, the balance shifts in favour of trickle-down and diffusion of modernity in all its material and non-material forms across the space economy. These processes were seen as essentially positive and beneficial, and the spatial inequalities (as well as the economic and social inequities which they represented) as a necessary price to pay in the course of development. Importantly, and explicitly, these were believed to be of limited duration, since the state of industrialised development was characterised by a high degree of territorial homogeneity and the elimination of the periphery and all that it represented.

Little problem was perceived in translating these theoretical propositions and models to Third World contexts, and no account was taken of

STAGE 1 Pre-industrial: Isolated, independent centres, with little interaction

STAGE 2 Transitional: Emergence of a dominant core

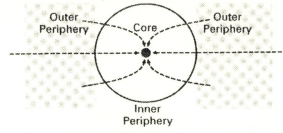

STAGE 3 Industrial: A single dominant centre with strong peripheral sub-cores in a shrinking and fragmented periphery

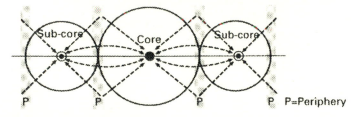

STAGE 4 Post-industrial: Fully integrated urban system which has virtually eliminated this periphery

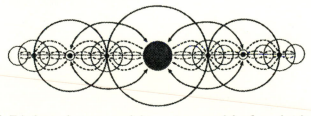

Figure 3.1 Friedmann's core–periphery stage model of modernisation and development
Source: After Friedmann, J. (1966) *Regional Development Policy: a case study of Venezuela*, Cambridge, Mass.: MIT Press

the very great differences in situation. A widely cited example of this, which also illustrates the point well, is Ed Taaffe, Richard Morrill and Peter Gould's (1963) model of transport network development, based on research in West Africa (Case study A). The successive stages are distinguished graphically and relate to the gradual expansion and development of colonial economies, ultimately providing the infrastructure inherited at independence. In a similar vein, James Vance produced a five-stage mercantile (i.e. trade-driven) model of the establishment and expansion of settlement patterns in a new colony and its colonising country. An important element in the model is the development of a transport network in the colony and the intensification of links and traffic between metropole and colony, reducing the effective distance between them.

Case study A

The evolution of transport networks in West Africa: a modernisation approach

The model published by Taaffe, Morrill and Gould (1963) exemplifies the modernisation theoretical approach to transport network evolution in a Third World context. It was based on historical research on the coastal countries of West Africa. Although it comprises six distinct phases, as represented in Figure A.1, in a manner comparable to Rostow's stages of growth, the authors allowed greater flexibility and realism by stating explicitly that not all countries or colonies would necessarily experience all the stages, nor need the stages be distinct or strictly sequential in practice.

Phase 1 comprises the establishment of a series of separate trading posts and minor ports along the coast by incoming merchants. These act as points of interaction where locally produced or traded goods are exchanged for imports (in this case from Europe) or money. In the second phase, trading routes begin to penetrate the hinterland from the coast – represented schematically in the diagram as a single line from each of the ports involved (i.e. P_1 and P_2). This enlarges the catchment of those ports and brings the traders into contact with new groups of people and different resources located at I_1 and I_2. Phase 3 is characterised by the extension of these hinterlands through the construction of lateral feeder routes from the trunk links. Each of the growing ports still has its own distinct network and hinterland, though.

Case study A *(continued)*

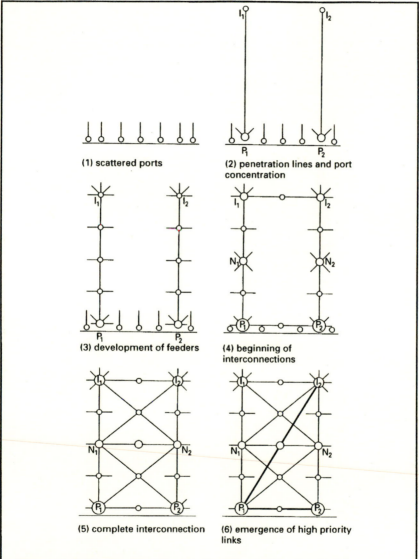

Figure A.1 Taaffe, Morrill and Gould's model of colonial transport network evolution
Source: After Taaffe, Morrill and Gould (1963)

Case study A *(continued)*

The fourth phase marks the beginning of interconnection between these different local/regional systems and the emergence of intermediate urban centres (N_1 and N_2). As economic development proceeds and new resources, e.g. mineral deposits or large-scale plantations, are exploited or transport interchanges and administrative centres are required, such new nodes in the interior develop. Although in specific cases, these may grow to rival the port city as the dominant centre, the ports have a distinct advantage. However, the forging of an increasingly integrated network (Phase 5) reduces the need for so many ports, with the result that a process of selection commences, whereby trade is increasingly concentrated in one or two large ports, either the largest city or the one with the most favourable port facilities and location. Most, if not all, of the small ports serving local hinterlands decline further or disappear altogether. Connectivity, which is defined as the extent to which each significant node or settlement is connected directly with the others, increases rapidly. The final, sixth phase resembles what has happened in rail networks in Northern countries like the UK and USA, namely the closure of branch and low-volume lines in favour of the enhancement of a more limited number of high-volume, high-priority links between the principal centres.

Critique

Such studies are presented in an apparently unproblematic and value-free manner by their authors. The construction and expansion of these networks is taken (implicitly and in some respects explicitly) to be undeniably positive. This fits with the notion that transport is an essential prerequisite for development. No account is taken of the fundamentally unequal power relations between colonists and colonised; after all, the West African coast and interior were densely populated by sophisticated indigenous societies long before the arrival of Europeans. How did colonisation and the advent of the railways (or other introduced technologies) affect these societies, and their social, political, economic and spatial organisation? The development of colonial infrastructure reflected the priorities and preferences of the colonists or imperial power, often with little regard for indigenes. In some areas, traditional modes and forms of transport were undermined. However, many important indigenous towns were bypassed by the new rail and road networks, disadvantaging them in terms of commercial activity or administrative functions. As a result such centres often retained a more indigenous character than places adopted or adapted by the new rulers. Either way, the effects were certainly not neutral.

Modernisation theory can also be criticised with reference to empirical data, for example on trends in private motor vehicle ownership (Case study B).

Case study B

Patterns of motor vehicle ownership

In terms of modernisation theory, one would expect a clear and roughly linear relationship to exist between a country's GNP per capita and levels of motor vehicle ownership. However, the most recent data available show the actual situation to be far more complex. Although at a highly generalised level it is possible to say that high-income countries tend to have higher ownership rates than low-income countries, there are substantial differences within and between income categories and continental regions (Table B.1). Second, within individual countries, one would expect ownership rates to rise as per capita incomes rise. This is generally the case, albeit with a few exceptions.

Let us examine the evidence on cross-country comparisons in Table B.1. The highest income countries are Switzerland, Japan, Denmark, the USA, France, Great Britain and Australia in descending order. However, the USA has by far the highest car ownership rate, followed by Australia, Switzerland, then very closely by France and Great Britain some way behind it. Japan's rate is even lower, despite its substantially higher GNP per capita. Conversely, Australia's figures are far higher than those for Great Britain, despite the latter enjoying a per capita GDP of over $500 more.

Table B.1 Registered cars, motorcycles and mopeds in use (selected countries)

Country	Year	Passenger cars	Motorcycles and mopeds	Population (thousands)	GNP per capita 1992 ($)	Number of cars per 1,000 people	Number of motorcycles and mopeds per 1,000 people
Europe							
Denmark	1989	1,591,546[1]	43,255	5,135		310	8.42
	1993	1,674,939[1]	47,405	5,181	26,000	323	9.15
France	1989	23,010,000	–	56,300		409	–
	1993	24,385,000	2,990,000	57,800	22,260	422	51.73
Great Britain	1989	19,266,000	875,000	55,700[2]		345	15.71
	1993	20,344,000	713,000	56,200	17,790	362	12.69
Greece	1989	1,605,181	219,547[3]	10,039		159.9	21.87
	1993	1,958,544	387,877[3]	10,390	7,290	188.5	37.33
Lithuania	1989	451,959	201,437	3,708		122	54.32
	1993	597,735	180,452	3,724	1,310	161	48.46
Poland	1989	4,846,411	1,410,859	38,042		127.4	37.09
	1993	6,770,557	1,067,634[3]	38,505	1,910	175.8	37.45
Romania	1989	1,218,128	294,590	23,151.6		52.6	12.72
	1993	1,793,054	322,891	22,751	1,130	78.8	14.19
Switzerland[4]	1989	2,916,959[5]	813,578	6,699		435.4	121.45
	1993	3,137,619	743,596	6,938	36,080	452.2	107.18
Africa							
Egypt	1989	1,054,465	330,521	56,100		18.8	5.89
	1992	1,119,727	351,496	59,167	640	18.9	5.94
Ethiopia	1989	39,942	1,708	48,696		0.8	0.04
	1991	37,799	1,515	52,000	110	0.73	0.03
Sierra Leone	1989	29,012	9,827	3,855		7.5	2.55
	1993	32,415	10,198	4,225	160	7.7	2.41
South Africa[6]	1989	3,316,706	303,251	29,908		110.9	10.14
	1992	3,522,129	285,034	31,917	2,670	110.4	8.93
Togo	1989	3,840	1,557	3,400		1.2	0.46
	1993	3,112	982	3,813	390	0.82	0.26
America							
Costa Rica	1989	143,860	40,620	2,958		48.6	13.73
	1993	220,142	46,000	3,167	1,960	69.5	14.52
Mexico	1990	6,839,337	249,722	81,250		84.2	3.07
	1992	7,497,128	263,568	84,500	3,470	88.7	3.12
Nicaragua	1989	39,771	12,306	3,574		11.13	3.44
	1992	67,158	22,221	3,969	340	16.92	5.60
USA	1989	143,025,658	4,433,915[7]	248,491		575.8	17.84
	1992	144,213,429	4,065,118[7]	225,078	23,240	565.4	18.06
Asia and Middle East							
Cambodia	1990	3,550	72,392	8,570		0.41	8.45
	1993	28,919	292,830	9,020	? but < 675	3.21	32.46
Hong Kong	1989	219,579[8]	18,829	5,812		37.8	3.24
	1993	315,552	28,066	6,020	15,360[9]	52.4	4.66
Japan	1989	32,621,085	18,207,552[10]	123,340		264.5	147.62
	1993	40,772,407	16,395,506[10]	123,788	28,190	329.4	132.45

Table B.1 continued

Korea, Republic of	1989	1,558,660	1,187,777	42,449		36.7	27.98
	1993	4,271,253	1,936,345	44,056	6,790	97.0	43.95
Pakistan	1989	395,672	818,398	108,678		3.64	7.53
	1993	554,907	1,166,493	112,801	420	4.52	10.34
Saudia Arabia	1989	2,550,465	12,204	11,514		221.5	1.06
	1991	2,762,132	13,038	11,861	7,510	232.8	1.10
Thailand	1989	656,169	4,138,608	55,888		12	74.05
	1993	1,022,137	7,106,893	58,336,072	1,840	17.5	0.12
Oceania							
Australia	1989	7,442,200[11]	316,600	16,833		443	18.81
	1993	8,050,000[11]	291,700	17,662	17,260	455.7	16.52

Notes:
[1] Including vans < 2 tons
[2] 1 July estimate
[3] Motorcycles only
[4] 30 September
[5] Includes station-wagons
[6] End of June
[7] Motorcycles only
[8] Including crown vehicles
[9] GDP (not GNP)
[10] As per 1 April
[11] Includes station-wagons

Source: derived from International Road Federation (1994) *World Road Statistics 1989–1993*, Geneva and Washington, DC: IRF

Case study B *(continued)*

Again, Greece, Poland and Lithuania have comparable car ownership levels despite their substantial income differences, while Romania has an ownership level comparable to many poor countries in the South. Conversely, Saudi Arabia, a relatively high-income oil-exporting country, has a car ownership rate substantially higher than Greece and over four times as high as Hong Kong, which has double Saudi Arabia's average GNP. Hong Kong also has a far lower rate than South Korea, Mexico and South Africa, three NICs or proto-NICs with substantially lower per capita GNP. At the same time, we know that the distribution of income in South Korea is far more equitable than in either Mexico or South Africa, where it is notoriously skewed.

Case study B *(continued)*

Among low-income countries, there is also no direct relationship between per capita GNP and ownership levels, although Ethiopia does have a substantially lower rate than even Sierra Leone. Perhaps surprisingly, Nicaragua's ownership rate is well above that of Pakistan, despite the latter having a per capita GNP some $80 higher.

Although the relationships between motorcycles and mopeds and per capita GNP are different, similar complexities exist. There is also no simple relationship between car and motorcycle ownership within individual countries. The expectation of higher motorcycle ownership than car ownership in the less poor low-income countries like Egypt and Pakistan and lower middle-income countries like Thailand, Costa Rica and Romania is no more consistently borne out by the data than the notion that ownership levels of motorbikes tend to fall in high-income countries.

These trends and complexities show clearly that per capita income is only one of the factors determining ownership rates of different types of private motor vehicle. Country size; quality of infrastructure; the availability, quality of service and affordability of public transport; cultural, religious and social values; state policies; and the distribution of incomes within countries are all important influences of which due account must be taken.

Challenging the orthodoxy:
political economy and dependency

In reaction both to perceived theoretical inadequacies of the modernisation approach and considerable empirical evidence of the failure of market-led economic development to transform Third World economies and societies, a number of critiques and alternative perspectives based on Marxist principles of political economy arose. Although the more extreme versions also emphasise the primary importance of the economy in determining human motivation and action, the basic assumption is precisely the opposite. Whereas neoclassical economics assumes an essentially benign, socially optimal outcome to perfect competition in the market, Marxist political economy is founded on the assumption that capitalism is characterised by class-based conflict between the owners of capital and the workers, and that outcomes are socially suboptimal. State intervention, as in social democracy, can do little more than ameliorate the worst excesses and hardships, since capitalism remains the underlying mode of production. According to such perspectives, the only solution is for capitalism to be overthrown in favour of socialism, or literally the 'dictatorship of the proletariat'. The essence of political economy is that the 'political' and the 'economic' are inextricably interwoven: political institutions and practices legitimise and serve the interests of those who hold power. The division between them which characterises Western systems and finds expression in academic disciplinary boundaries is both artificial and a device for perpetuating the myth that economics is (or should be) governed by (free) market mechanisms rather than by politically determined legislation.

Extreme Marxist formulations have never gained significant favour in the West, and since the collapse of the former Second World (the USSR and its allies) in the late 1980s, even more moderate political economy ideas have been widely discredited as unrealistic. This applies particularly to the field of Transport Studies, where political economy has to date had very little impact on mainstream thought, in either academic or practitioner circles.

Probably the most widely known school of thought within this tradition is 'dependency theory'. Formulated in the late 1960s and early 1970s, it draws on neo-Marxist principles, seeking to reinterpret classical Marxism in a more contemporarily relevant manner. Essentially, the *dependistas* argue that the Third World is held in an unequal and exploitative relationship by the forces of global capitalism, symbolised by transnational corporations, which are based in the metropolitan cores of the North. The form which capitalism takes in the periphery is a distorted and subordinate form of metropolitan capitalism, lacking the potential or dynamism to develop any autonomy. The Third World's subordination ensures that surplus value will continue to be extracted there and transferred, via unbalanced trade and inequitable pricing mechanisms, to the metropoles. Any industrialisation occurring in the South is of the enclave type, geared to capital accumulation in Europe and North America rather than having the potential to act as an agent of generalised development outside the urban centres. Although not explicitly addressed within the paradigm, transport again forms the key bridge between core and periphery. While the technologies in themselves may be neutral, they were developed in the North under particular politico-economic and technological conditions, and are controlled by the North and utilised to serve those interests.

The significance of this perspective was, first, that it arose in Latin America rather than Europe or North America, the birthplace of most mainstream theories. As such it formed part of a Third World critique of the Northern monopoly of intellectual agendas and views of the world. Second, it was grounded in the perceived failures of modernisation and even the more progressive, import-substituting industrialisation programmes implemented in Latin America since the mid-1950s. Although commonly known as 'dependency theory', it is not a true theory, lacking causative explanation of how the current situation arose and a predictive element capable of dealing with change. It could not explain the rise of the NICs, the evolution of transnational corporations headquartered in the South, or the growing diversity of conditions within the Third World.

Nevertheless, it had a marked impact on social-scientific thought and rightly focused our attention on the truly *global* scale of politico-economic relations and the interconnection between forces and processes in metropolitan and (ex-)colonial countries. It is also overly deterministic, because the pure or extreme versions, as elaborated most notably by André Gunder Frank in the Latin American context and by Samir Amin and Walter Rodney with respect to Africa, argued that there was little that dependent peripheral countries could do about their plight. Action had to be taken at a global level (a very tall order indeed) in view of the global nature of the problem. The only option open to an individual country seeking to free itself of such exploitative relations was to cut its ties with the world economy, in other words to 'go it alone' in an isolationist strategy (sometimes known as autarchy). To deny the validity of analysis and policy at the national level is unrealistic, for national governments remain the principal level of political decision-making.

World-systems theory, as expounded over the last twenty years by Immanuel Wallerstein, represents a refinement and elaboration of the essential theses of dependency, dividing the world into three broad categories of economic development (core, semi-periphery and periphery) in order to include the NICs in the schema, reducing the determinism and taking greater account of the role of different state forms in various periods over the last 500 years. Another strand of theory is known as structural Marxism, which emphasises important elements in classical Marxism and focuses on the operation and role of structural and institutional factors in determining outcomes and constraining alternative development paths.

In the context of transport, David Slater (1975) showed how the development of colonial transport networks reflected colonists' perceptions of strategic and economic value in conquered territories, thus serving the interests of the colonial power directly. Moreover, these networks provided the means whereby, and partly determined the extent to which, different regions were incorporated into the world-economy. In Tanganyika (now mainland Tanzania), for example, the territorial space economy had already been divided into three distinct economic zones by the 1930s. The coastal belt, comprising the major towns, sisal plantations and other cash crop-producing regions, specialising in production for export, were the most directly linked to the transport network. A second zone, surrounding the first, supplied the latter with food and services, while the third and most peripheral zone served as a source of migrant labour from stagnant or declining subsistence peasant economies.

Although this conceptualisation reproduces a threefold core–semi-periphery–periphery structure, the underlying theoretical propositions are substantially different, emphasising the rationale for and nature of Tanganyika's insertion into the world-economy, and making explicit how this often cut across existing precolonial modes of production, political and administrative structures and transport routes. This Slater characterised as simultaneous processes of internal disintegration and increasing external integration. A similar correspondence between the extent and quality of the transport infrastructure and zones of colonial importance is evident in many former colonies (cf. Ezeife and Bolade 1984), and reflects the network evolution model of Taaffe et al. (1963), although, as discussed above, the latter is conceptualised in a rather different way.

At independence, the new states faced many formidable challenges, given their frequently inadequate infrastructural and revenue bases. The colonial infrastructural inheritance, especially of railway networks, had been designed to serve the interests of political rulers thousands of kilometres away rather than the needs of the indigenous population. Many nationalist movements won power at independence on a platform of addressing local needs and distributing 'development' more equitably across the population and national space. The trajectories pursued by these governments can be divided into two broad categories: overtly capitalist, which sought to enhance the productive base and efficiency of the existing roles of those countries as producers of (mainly) primary commodities for the world market by modernising and industrialising; and some form of more radical, often 'socialist' transformation designed to emphasise local needs and priorities ahead of the dictates of the world market.

Especially for governments attempting the latter course, like that of Julius Nyerere in Tanzania, the primary transport infrastructure presented a formidable hurdle, in that it could not easily be relocated and rerouted to link different regions of the country hitherto deemed unimportant. Yet precisely such a network would be required to give effect to the new policies and programmes. Foreign donor funds enabled the construction of trunk roads and secondary networks in various regions as part of regional or integrated rural development programmes, although little change has occurred in the inherited rail system of three roughly parallel and unlinked lines running broadly east–west to the main ports. This can hardly be called a network. Tanzania is by no means unusual in this respect, as a glance at the map of Africa or Latin America will show. Virtually all the railways built in Africa since independence (i.e. the last 30–40 years) have actually continued the colonial pattern of linking new mines or other points of bulk resource extraction to the nearest port. This emphasises the persistence of neocolonial relations, whereby resources are still generally exported to the North in raw or semi-processed form.

A few other geographers have applied the principles of political economy to the transport sector. The particular value of such approaches is to indicate the underlying processes and forces, the negative as well as merely the positive impacts of technological change and the extension of colonial control. Political economy is particularly well suited to analysis of impacts upon different groups of people defined in terms of their degree of access to and control over the means of production, especially capital itself. The focus is on indigenous as well as settler communities and groups, on differentiation and exploitation as well as enrichment and progress, on underdevelopment as well as 'development'. This can be done at different scales, highlighting the territorial or spatial effects alongside the social. Another instructive example relating to the introduction of railways – this time in southern Africa – is provided in Case study C. The work of Mike McCall in the context of rural development will be discussed in Chapter 4.

Case study C

The political economy of railways in southern Africa

An excellent illustration of the profound and rapid effects of connection to the expanding rail network is provided by the immediate transformation of diamond mining in Kimberley in South Africa's Northern Cape province. Initial workings from 1869 onwards were small scale and alluvial, and a process of gradual agglomeration commenced very soon after. However, with the completion of the railway lines from the ports in the early 1880s, it became possible to import heavier steam-driven and then electric machinery, which in turn facilitated larger scale, deeper and more profitable excavation. The concentration of ownership in progressively fewer hands accelerated almost immediately, as smaller workings were bought up. The control and subordination of labour also became easier as a result of this, leading to the introduction of the barrack-like compound system for black workers and the origins of the migrant labour system, which later became institutionalised across southern Africa and beyond. Ultimately, De Beers was to emerge as an effective monopoly mining company.

Across the subcontinent, railways contributed to underdevelopment and poverty in four main ways, according to Gordon Pirie (1982):

1   They induced indebtedness because of the imperial requirement that colonies be self-financing and the inability of small or poor territories to fund and repay the costs of railway construction. Nyasaland (now Malawi) was forced by unilateral British decision to guarantee the interest and debentures for the Trans-Zambezia Railway, a project of far greater imperial political importance than its economic value to Nyasaland. As the line was a commercial failure, this obligation continued to drain the protectorate's coffers for many years.

2   They promoted the establishment of dependency relations, both on overseas sources of finance and equipment and maintenance, and by smaller, landlocked territories like Northern and Southern Rhodesia (now Zambia and Zimbabwe), Nyasaland and

Case study C *(continued)*

Swaziland on coastal neighbours, especially South Africa and Mozambique.

3 They accelerated proletarianisation and the extension of wage labour. Construction and operation of the railways employed large numbers of people, often working under extremely harsh conditions in the late nineteenth and early twentieth centuries. Equally, the railways quickly supplanted most transport riding by ox-wagon and horseback, which had hitherto been the major forms of long-distance transport. Many of the people employed by these small, privately owned and operated businesses had been Africans, often peasants working part-time. Loss of these jobs was not compensated for by new business or managerial opportunities on the railways. Railways were also utilised as an element of colonial policy to promote European settler agriculture, through the provision of local railheads, at the expense of sometimes successful African peasants, who were deprived of such access in newly established reserves.

4 They facilitated the extension and entrenchment of long-distance migrant labour across the subcontinent, by providing quick, reliable and cheap transport links between the gold, diamond and coal mining areas of South Africa and new recruitment areas in neighbouring countries as far away as Tanganyika (now Tanzania).

The above examples illustrate well that, while the introduction of transport networks and new transport technologies may promote and be necessary for development, it neither leads automatically and inevitably to development nor is necessarily an unequivocal blessing. The impacts are likely to be differential. More importantly, the same technological innovation may have very different impacts in different contexts, depending on local conditions, social formations and relations of production. The vastly different power relations and other conditions pertaining in various tropical and subtropical regions colonised by Europe suggest the need for careful analysis of those contexts rather than merely extrapolating the excitement and wonder of new

Northern technologies, and their supposedly undoubted benefits, to the (ex-)colonies.

Another danger implicit in such ethnocentric or Eurocentric assumptions is that of technological imperialism. It is very easy to assume that new, 'superior' technologies are developed in more 'advanced' societies, that this societal hierarchy is innate or 'natural', and has persisted indefinitely as a result of racial or environmental superiority. In his major recent book challenging the Eurocentric view of colonialism and geographical diffusionism, Jim Blaut specifically warns that:

> technological determinism gets its greatest strength from the error known as 'telescoping history.' When we travel, mentally, back to medieval Europe, we pass backward through the eras in which Europeans clearly were technologically superior to everyone else, and so we tend to expect that superiority to have been the state of things at all prior times. But Europe advanced technologically beyond Asia and Africa mainly after the beginning of the industrial revolution. Europe did not even begin to forge ahead of other civilizations in technology or science until the seventeenth century or even later. By telescoping history we imbue the Middle Ages with the marvelous technological attributes of modern Europe. It is then but a small step to the conclusion that, since technology is so *obviously* a powerful cause in history, and since Europe has *always* been so technological, *here* is the root of the European miracle.
>
> But tools do not invent themselves or reproduce themselves. If you invoke technological determinism, you must not only show that the technology appeared and had such-and-such effects, you must also explain why it was invented, and by whom. In nearly all (I think) technologically deterministic arguments made as part of some explanation for Europe's historical progress, technological arguments end up being arguments about the inventors, not the inventions. They end up in some Weberian claim that Europeans are more inventive, innovative, 'rational,' than non-Europeans. Technological determinism then differs from other kinds of tunnel-historical theories mainly in its claim that rational Europeans moved their society forward by inventing new technology, rather than doing so by inventing new political systems, new forms of social organization, new religions, or whatever.

(Blaut 1993: 108–9)

Because of its central concern with introducing and diffusing 'modern' technologies and processes, modernisation theory lends itself directly to such deterministic arguments, and its adherents have generally made little attempt either to make such assumptions explicit or to reject them. Conversely, by reacting fundamentally against the unequal exchange and exploitation of colonialism and neocolonialism (i.e. the persistence or maintenance of colonial-type relations after political independence), many *dependistas* and other critics in the neo-Marxist tradition have tended to idealise precolonial and other indigenous political, economic and social forms and technologies, seeing the railway and motor car almost as symbols of evil and to be rejected (even if they use them themselves!). This is clearly no more helpful than denying the scope for positive action at the level of the nation state. One cannot go back in time, undo history or remove existing technologies without finding a superior and acceptable replacement.

The postmodern critique

In the last decade or so, an increasingly influential intellectual movement, known as postmodernism, has mounted a fundamental critique of all the types of theories and ways of theorising outlined above, on the grounds that they themselves typify modernism. This latter is the movement which – be it in literature, art, architecture or the social sciences – legitimised the 'modern' era and its preoccupation with modernisation in technology, social and political organisation. The nation state is also regarded as one of the increasingly outmoded and anachronistic inheritances of this era, nowhere more so than in the Third World, where this political form was imposed by European colonialism and imperialism. The critique of modernist theory-building argues that the modernisation and political economy traditions share an unhelpful logic and way of arguing and theorising, namely their preoccupation with seeking 'the truth', 'the best way', as if there were somehow a single theory that could surpass all others and have a monopoly of the truth. Moreover, these theories were then extrapolated from the specific (usually Northern) context in which they were developed and asserted to have universal relevance and applicability – hence the term 'universalising theories'.

Although postmodernism is a very heterogeneous movement, with diverse strands and schools, both in different disciplines or subject areas and within them, central features of the approach are:

- the 'decentring' (i.e. downgrading, replacement) of universalising theories and methodologies;
- concern instead with a multiplicity of more locally relevant and sensitive perspectives, insights and theories; in other words, various forms of pluralism;
- downgrading of the primary role accorded in theory and political practice to the nation state, in favour of more local and responsive forms of social and political organisation;
- replacement of monolithic styles in art, architecture and the like with diversity and mixtures ('pastiche') to reflect the different historical periods, the diverse cultures and social groups within almost all societies today.

In extreme forms, the thrust of postmodernism is totally anti-state and virtually anarchic, in that everything is seen as relative, as all right. Postmodernism is also expressed in different ways: as 'after' the modern, as anti-modern, and as an alternative to the modern. This has particular importance in the context of the South, where many parts of many countries and many of the population, especially in rural but even in major urban areas, have yet to experience substantive modernity. Moreover, as discussed earlier, the central policy thrust of most governments remains, whatever the particular rhetoric, the promotion of modernisation. The debt crisis and impact of structural adjustment have intensified these pressures and preoccupations.

There has therefore been academic debate on whether postmodernism can have relevance in such situations. The answer clearly depends on the interpretation of postmodernism: if seen as 'after' the modern, only a handful of countries, most conspicuously the NICs, could fit the description. However, in the senses of anti-modern or as an alternative to it, there is indeed relevance. Although a distinctively postmodern transport literature has yet to emerge, particular forms of transport, perhaps most conspicuously the wide-bodied jet and ocean cruise liner, have played a central role in promoting intercontinental package tourism. This is one of the fastest growing service industries, elevated to a central position in the development plans of many Third World countries. It is also an industry in which globalised patterns of consumption, the emergence of idealised, sanitised versions of local history and pastiche forms of hybridisation in architectural style have been most evident in recent years.

Other theoretical schools have emerged alongside, and to some extent intertwined with, postmodernism, namely postcolonialism and postdevelopmentalism. The former is concerned primarily with the recovery of histories and identities 'lost' during the colonial period and subsequently through the continued use of theories and perspectives derived during that experience and serving to legitimise it. Hence attention is focused particularly on subordinate and marginalised groups, including women, minorities and others who feel or felt voiceless. Many of the leading lights in this movement have been from the Third World.

Postdevelopmentalism shares an important intellectual position with postmodernism, arguing that the whole notion of 'developmentalism', namely that ex-colonies can and should be developed through concerted action directed predominantly from the North and by national elites in the South, is itself an essentially inappropriate and modernist conception. This certainly echoes the arguments of other opponents of Eurocentric notions and theories, but seeks to put Third World actors, especially the marginalised and peripheral groups, and their agendas centre stage. Less clear at present is how such agendas might be implementable or implemented without either resorting to autarchy or forging alliances with existing centres of power at local, national and international level which will ultimately be little different from those prevailing at present. The way forward would seem to be to derive less monolithic and dominant, more pluralistic and locally centred approaches which meet people's needs and aspirations. Such an approach would not be unique to postdevelopmentalism. Moreover, it should be remembered that most people *do* have aspirations and notions of 'development', however vague or even contradictory, in terms of where and how they see or would like to see themselves living at some point in the future.

Conclusions

It is easy to regard technology as the product of rational and innovative societies, and as being neutral in its effects. This would be a fundamental misunderstanding and oversimplification. In surveying and evaluating different theoretical perspectives on development, and their treatment of transport and technological innovation, this chapter has aimed to challenge conventional wisdom in this regard and to sensitise the reader to the importance of theories as guiding sets of ideas for 'seeing the world'.

As we have seen, development theories offer at best partial explanations of the processes and forces which operate; until very recently they also sought universal applicability. Modernisation and the various strands of political economy are products of the modern era, and their respective adherents have been concerned to establish the superiority of their version of 'the truth', their view of the world. Postmodern and postdevelopmental critiques have rejected the notion of a single path, promoting the case for heterogeneity and difference, for local relevance and control. Although postmodernists come from a very different direction, the thrust of these propositions may give renewed relevance to and opportunities for strong basic needs strategies and 'bottom-up' planning which shift the nature and balance of the relationships between the central and regional state and local communities.

The lessons of the discussions above are that transport issues are amenable to theoretical analysis and, moreover, that the assumptions and implications of any particular perspective should be made explicit. Modernisation theory has enjoyed extraordinary prominence and a virtually unchallenged status within the transport field. This reflects its appealing and optimistic logic, and the artificial distinction between 'positive' (supposedly value-free) and 'normative' (value-laden) approaches in neoclassical economics (upon which modernisation theory is based) that has allowed politicians, planners, consultants and academics to duck difficult issues in the name of a rational, value-free approach through which everyone will ultimately benefit as a result of diffusion or 'trickle-down'.

Political economy lends itself to more critical examination of conflict situations, and the nature of winners and losers. This was well illustrated with reference to the introduction of railways in Tanganyika and southern Africa.

Key ideas

1 There is a powerful myth of atheoretical or value-free research and planning in the transport field. In such cases, the value judgements and assumptions are implicit rather than explicit, but no less important or influential.
2 Modernisation theory has dominated in the transport field more completely and with less overt challenge than in most other disciplines in the social sciences. This perspective sees transport and technological innovation as important and beneficial to the process of (economic) development.

3 Various schools of thought under the 'political economy' umbrella have sought to challenge this view. Some such perspectives were too deterministic and rigid, but the value of a more moderate approach is its appropriateness to distinguishing the role of and impact on different social groups and regions of territory in terms of their relationship to the transport technology and how it is being utilised.

4 The existence or construction of a new transport link does not mean that development will automatically or necessarily follow. Previously self-reliant areas and communities can be undermined through being integrated into wider systems to which they are marginal and over which they have no effective control, for example, in determining the terms of trade.

5 Postmodern and other recent critiques have challenged the 'modernist' rationale of modernisation, political economy and other theories in seeking to universalise their applicability and establish themselves as superior to other explanations. Instead, a multiplicity of less 'grand' theories and explanations should be welcomed.

4
Rural transport, accessibility and development

Introduction

In many respects, the actual or potential impacts of transport policy and infrastructural provision on development are more readily examined in rural than urban contexts, and it is to this that attention now shifts in the light of the issues covered in Chapter 3.

It is well known that many rural dwellers experience very limited mobility. It is also widely accepted that – as a generalisation – a high proportion of people in rural areas of most Third World countries are poor, whether measured in absolute or relative terms (see Chapter 1). These two characteristics are often assumed, especially by those who adhere implicitly or explicitly to modernisation theory, to be causally related. In other words, the people concerned are thought to have low mobility *because* they are poor. Low mobility is also commonly assumed to be synonymous with inadequate mobility. This may be fair as a general assumption, but only very seldom have perception or attitudinal surveys of poor rural dwellers been undertaken to examine the question or, indeed, to rank their priorities as between improved accessibility or mobility and other development needs.

Be that as it may, the modernisation approach would then suggest that poverty could be tackled through interventions which improve access and mobility, either on their own or as part of wider, 'integrated' rural development programmes. This is precisely the rationale and justification for numerous investment programmes in rural regions of Third

World countries since the 1960s. These have proved particularly attractive to bilateral and multilateral donor organisations because they are very visible and are felt to have relatively rapid and tangible benefits. In addition, road and rail schemes often provide opportunities for civil engineering and other contractors from donor countries and perhaps even export orders for transport equipment. Under such conditions, especially if the aid can be 'tied', a significant proportion of the total resources flow back to the donor country or countries, providing an increasingly important economic justification to governments for giving official development assistance in the first place.

In practice, rural development schemes and infrastructural projects have had very mixed results. The degree of success has depended on numerous factors, of which the most important are appropriateness to local conditions (including intensity of use), integration with local institutions and priorities, and the ability of the recipient country to maintain the new infrastructure, in terms of skills, equipment and other resources, and institutional will. Generally, the simpler the project in terms of objectives, technology and scale, the greater its chances of success, although this does not mean that 'low' technology is inevitably 'best' or most appropriate. The Third World is littered with prestigious 'white elephants': impressive projects of all descriptions, including trunk and feeder roads and telecommunications systems, which began to deteriorate immediately after construction or as soon as the last expatriate staff left, because one or more of these conditions was not fulfilled. Some of these are monuments to political and engineering folly on the part of donors and/or the recipients; others simply to unfortunate circumstances and the lack of provision for a maintenance budget and capacity when the projects were implemented. We will return to the issues of technology and appropriateness in more detail below.

Debating rural access and mobility

It is certainly true that most peasant communities – which by definition produce food principally for their own consumption – are less mobile and may have lower access to areas outside their immediate vicinity than those who sell some or all of their output commercially. This reflects 'need' or the lack thereof, rather than any inherent state of 'underdevelopment' or poverty. Although, in comparison with commercial farmers, many peasants would be classified as poor, this does not necessarily mean that they are unhappy or wish for a radical change in lifestyle. In the past, many peasant communities in tropical and subtropical zones were well able to provide adequately for their own needs, and even grow a surplus, under broadly sustainable conditions. Moreover, 'traditional' pastoralist societies were much more mobile than peasants, often covering large distances in their patterns of transhumance (seasonal migration in accordance with grazing conditions and water availability).

Colonisation, modernisation and accessibility

Today transhumance is frequently not possible; the changes have been brought about by the impact of modernisation – usually as a consequence of European imperial contact or direct colonisation – and the undermining of self-sufficiency and self-reliance in a number of ways. Sometimes this took the form of land theft, the expropriation of high-quality land with good agricultural potential, for use by European settler farmers or the establishment of plantations. This reduced the land available to indigenous communities, often relegating them to less suitable agro-ecological zones and/or cutting off seasonal migration routes and water sources. This increased pressure on remaining land over time, setting in motion processes of progressive intensification and/ or land degradation and erosion, depending on local circumstances. Land held under communal tenure systems was particularly vulnerable to erosion under such conditions. In southern and parts of East Africa, land expropriations formed part of a deliberate strategy to coerce labour into the colonial economies under exploitative and demeaning conditions. Hut or poll taxes, payable in cash only, were imposed and, since land and marketing opportunities were inadequate, the most reliable way of raising the money was to undertake wage labour. In other regions, colonial taxes were used as a method of forcing indigenous communities to produce cash crops for export to Europe.

Although widespread, such measures were not universal, and even where peasant communities flourished through the ability to market surplus crops which the establishment of colonial towns and transport networks facilitated, these changes had far-reaching ramifications. These include increasing economic and social stratification, leading to disparities and new social and political tensions over land, other resources and power. A process of westernisation commenced, affecting different people within communities to differing extents. Generally those villages and areas closest to the emerging colonial economy, urban areas or transport routes experienced the most dramatic transformations, although migrant labour flows sometimes became most intense from distant regions where alternatives were lacking, e.g. from western Kenya to Nairobi, or from central Africa to the Transvaal gold mines and the Zambian Copperbelt.

Of course today virtually all former colonies are sovereign states. However, the colonial legacy often remains strong, in terms of both the very nature and boundaries of the nation state system, and the physical infrastructure, structure of the economy and, indeed, of society. Terminology like 'remote' and 'the need for development' is used by governments, planners and others at the centre to describe people and places which are very distant from larger centres or the nearest roads either physically or in terms of the time and effort needed to reach them. It is important, too, to realise that – as explained in Chapter 3 – facilities and infrastructure were usually situated according to colonial needs, not the needs, priorities or even geographical distribution of the indigenous population, unless the two coincided.

Generally, 'remote' areas and communities have lower degrees of westernisation, are poorer and have more limited mobility. However, it would be interesting to know whether such people necessarily considered themselves to be 'remote', cut off, or in need of 'development' in the past, even if they do today. The point is that our terminology is implicitly shaped by the political, economic and social centre, not the periphery. Yet supposedly remote places may be very central and important to alternative, indigenous social, political and economic entities or networks which bear little resemblance to modern ones and which have, in some form or another, survived to the present. There are numerous examples on every continent in the South where colonial borders form an artificial division across indigenous territories, spheres of influence and trade routes, so that communities on either side are linked far more profoundly by common identities and interests than they are divided by their citizenship of different countries. Such border areas are often the sites of conflict between states seeking to assert their control over all their territory and all interactions and transactions across their borders.

The state as provider or promoter of development

Border conflicts are actually an example of a wider phenomenon, namely the actions of developmentalist states, in terms of which the state at all levels sees itself as the prime agent and principal promoter of 'development'. As argued in Chapter 3, governments of newly independent countries have been faced with fundamental decisions over which ideology and direction of development to follow. The extreme positions were: to continue as before, only more efficiently and supposedly more equitably, serving the world market; or to attempt a reorientation and transformation to meet local needs first. In reality, most states have opted for some intermediate strategy including elements of both objectives. With a tiny handful of brutal exceptions, such as the Khmer Rouge in Cambodia (Kampuchea) or Idi Amin's regime in Uganda, Third World governments have not fundamentally challenged the modernist and developmentalist agenda of promoting 'development', irrespective of the guiding ideology invoked in the process.

Hence, development has been pursued by seeking to promote national integration, both physically and economically, and by forging a new, postcolonial national identity. Physical integration has involved the construction of new domestic airports, trunk roads and railways linking different regions, and the provision of lower order routes, including feeder roads, to rural communities hitherto bypassed by the transport networks. The development plans of many countries have referred to such integration as essential for development, in terms both of providing access to markets and facilitating the construction and provision of other development infrastructure, e.g. schools, hospitals, clinics and agricultural extension services, by the state. Whereas high-order transport investments are often undertaken separately, as with trunk roads or airports, more local roads are nowadays frequently linked to wider, multifaceted programmes designed to promote rural development and funded partly or entirely by foreign donors.

Road construction as a route to development

In many respects, the construction of roads has facilitated greater interaction, institutional development and integration with the rest of the country concerned. However, experience has shown that significant problems or side effects may arise too. It is therefore important to bear several qualifications in mind:

1 'Development', however framed, has not inevitably followed the opening of a new road. If there are no complementary resources or products to form the basis for trade, if the demand for travel is low,

or if other constraints outweigh the value of the new transport connection, the link may become an under-utilised 'white elephant'.

2   Incorporation into networks can actually undermine local production and skills in 'remote' or peripheral areas by facilitating the large-scale import of externally produced goods and foodstuffs; the activities of traders buying local produce at artificially low prices because of poor information, lack of alternatives and/or their effective monopoly power; and the outmigration of young, relatively skilled or able-bodied inhabitants.

3   It is common for new roads and transport technologies to have differential impacts, both positive and negative, which are expressed spatially, economically and socially. Space should, of course, be conceived here as a relative variable, serving as a differentiator of social processes rather than as a separate entity. The differential impacts occur at different scales, between regions or zones, as explained in Chapter 3; between localities, as will be exemplified below, and between groups of people in the same locality.

4   Undeclared motives behind infrastructural expansion may be just as important as those stated in development plans or project documents. For example, when new rural areas are included in road and communications networks, the state is better able to exert its control there, to impose its political will and to raise taxes. This relates to the point made earlier about control of borders and territory. In several countries, critical analysts have perceived these political and economic objectives as crucial, with development being used expediently as a form of legitimisation for extending authoritarian control and enhancing levels of production in order to increase tax revenue. In this light, Tanzania's *ujamaa* villagisation programme of the 1970s, which became increasingly coercive over time and to which a good road network was essential, has been criticised by some as being primarily a tool of the state for 'capturing the peasantry'. In other words, the balance of impacts may be negative or exploitative as well as positive.

5   New roads may give rise to profound environmental and health problems, for example soil erosion if sloping soils are exposed in areas of heavy rainfall, enabling unsustainable logging and mineral prospecting and mining in previously undisturbed forests, as in Amazonia, or the introduction of previously unknown diseases to which local people have little immunity.

Most of these issues are not considered in the rather narrow cost–benefit analyses in terms of which most road construction decisions and priority rankings have been formulated and justified. There has been a move over the last decade or so, at least by major development agencies, to take greater account of environmental impacts, although this is still done by attaching theoretical monetary values to certain environmental assets rather than taking on board any wider, more fundamental considerations.

Trunk routes

Trunk routes link different (sub-)regions of a country, serving as principal connections and carrying high traffic volumes. Nowadays, they are usually designed to follow the shortest route to their destination, in order to minimise construction costs and subsequent user travel times. They are generally of high quality and are designed to carry large volumes of heavy traffic (Plate 4.1). The directness of trunk routes has been facilitated by more sophisticated engineering skills, enabling tunnels through mountains to replace long, winding passes, or the spanning of river gorges with impressive bridges which can be

Plate 4.1 The high-quality national road north of Okahandja, Namibia

used at speed so that vehicles do not have to slow down and ford the actual river. The shortest route is therefore not necessarily the cheapest to build, although it will generate travel cost savings in the economy afterwards, may reduce accidents by being a safer design and may be environmentally less damaging. For these reasons, parts of existing trunk routes are sometimes realigned, where traffic volumes, accident rates and/or deterioration of the existing surface justify it. An instructive example of this and the various impacts, is provided in Case study D. As with all roads, maintenance is very important, and the integrity of many aid-funded tarred highways has been undermined through inadequate attention to this (Plate 4.2).

Case study D

Realignment of a trunk road in Nigeria

The realignment of a portion of the main road linking Zaria and Kano, the two principal cities in northern Nigeria, illustrates well the complex development impacts of such transport investments. The new 40-kilometre stretch immediately north of Zaria, completed in 1978, shortened the distance between the cities by some 20 km. It is far straighter, running parallel to the railway line, and is situated some 10–15 km east of the old alignment.

Research conducted along both routes in 1981 found marked differences in perceptions and opportunities. Employment and some agricultural opportunities along the old road were declining, while the new road had opened up many new ones as a result of the potential market represented by the passing traffic and because of better access to Zaria city. This was reflected in considerable migration flows to a corridor straddling the new road from the area adjacent to the old road and from further east (Figure D.1). Over 40 per cent of the 257 households interviewed along the old road had lost at least one member through such migration over the intervening three years, the majority of them male traders to whom the road was vital. Conversely, over half the 249 respondents along the new road had moved there since its completion, although many of these came from Zaria, not the vicinity of the old road.

Plate 4.2 The tarred bridge embankment of a trunk road undermined by heavy rain, Namibia. If not repaired quickly, the entire bridge could become unusable

Case study D *(continued)*

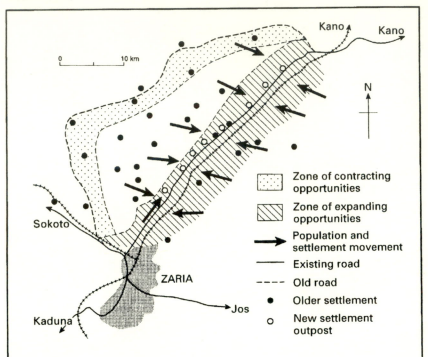

Figure D.1 Population shifts along the old and new Zaria–Kano road sections
Source: Modified from Salau and Baba (1984), Figures 2 and 3

This substantial population shift was having a marked effect on land values: although general price inflation was substantial over this period, the rate of increase was far more significant and also more widely perceived near the new road. Agricultural land use also changed as a result of land values and especially the marketing opportunities along the new road. Here tomatoes, vegetables, potatoes and other crops easily sold to motorists were most common, whereas along the old road, such crops had declined in favour of maize, rice and sugar cane.

Respondents in both areas agreed that improved accessibility was the most important benefit of the new road (Figure D.2),

Case study D *(continued)*

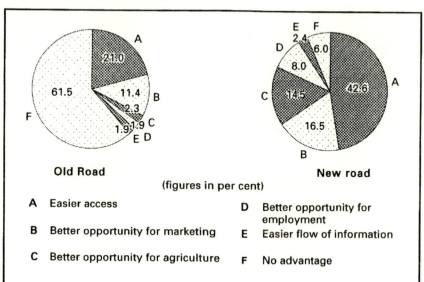

Old Road   **New road**

(figures in per cent)

A   Easier access   D   Better opportunity for
                                                     employment

B   Better opportunity for marketing   E   Easier flow of information

C   Better opportunity for agriculture   F   No advantage

Figure D.2 Major perceived advantages of new Zaria–Kano road alignment
Source: Based on data in Salau and Baba (1984)

followed by better opportunities for marketing and agriculture respectively. However, the proportions of respondents citing these advantages were far higher along the new road, and comprised three-quarters of all respondents there. Conversely, these three responses accounted for only 35 per cent of interviewees along the old road, where over 61 per cent felt that there was no advantage at all. In a supplementary question, over 93 per cent of respondents along the new road felt that movement had been facilitated to a small or large extent, compared with only 38 per cent along the old road.

It is therefore unsurprising that very different changes in visiting patterns to Zaria emerged between the two areas (Table D.1). Fewer people along the old road now made the trip daily, while the weekly, monthly and annual figures have remained constant; the only increase was in the category of 'other frequency'. By contrast, daily and weekly trips from the new road area had risen markedly, whereas monthly and annual visits had declined equally

Case study D *(continued)*

Table D.1 Frequency of visits to Zaria before and after road realignment

| | Old Road | | | | New Road | | | |
| | Before | | After | | Before | | After | |
	No.	%	No.	%	No.	%	No.	%
Every day	46	17.9	29	11.3	8	3.2	25	10.0
Once a week	96	37.4	98	38.1	29	11.6	93	37.3
Once a month	82	31.9	81	31.5	106	42.6	44	17.7
Once a year	12	4.7	9	3.5	62	24.9	32	12.9
Other frequency	13	5.1	25	9.7	23	9.2	31	12.4
Not sure	8	3.1	15	5.8	21	8.4	24	9.6
Total	257	100.0	257	100.0	249	100.0	249	100.0

Source: Salau, A.T. and Baba, J.M. (1984)

sharply. The principal reasons cited by both groups for such visits were to buy and sell goods and to visit family or friends.

It should be pointed out that many respondents felt that new local market roads would have been more helpful than the trunk road, in order to facilitate their access to rural (mainly periodic) markets, which were closer to home than Zaria. However, there had clearly been considerable switching of sale venue from periodic markets and Zaria to the new roadside.

Source: Salau, A.T. and Baba, J.M. (1984) 'The spatial impact of the relocation of a section of the Zaria–Kano road: a study of change and development in rural Zaria', *Applied Geography* 4(4): 283–92.

The region where the construction of major new trunk roads in the name of development has had the most dramatic impact is the Brazilian Amazon. This vast and hitherto remote and inaccessible area – in relation to the rest of Brazil – has been 'opened up' to settlement by poor, landless people from other regions, commercial logging on a vast scale, iron ore mining and even cattle ranching, often with rapid and dramatic effect. The environment is paying a high price, and has become something of a *cause célèbre* for international environmental movements, acting as a focus for concern about the build-up of so-called greenhouse gases in the atmosphere and hence the problem of global warming.

The Polonoroeste (Northwest Pole) highway programme in Rondonia Province attracted notoriety in the late 1980s on account of the extent of environmental damage caused during construction, as well as the wider impact of large-scale legal and illegal forest clearance triggered along its route. Tens of thousands of landless settlers, with no farming experience or none relevant to the local conditions, were encouraged to settle in the area and commence farming. This became one of the world's last great land frontiers, providing a means for the Brazilian government to assuage land hunger and defuse the political pressure building up in the country's southeast heartland as a result. Other forest land was cleared for large commercial cattle ranches, despite the unsuitability of the tropical soils to such activities.

Some interesting research has also been undertaken into the relationship between new trunk road construction and health conditions. Airey (1989) studied the effects of road improvements on patient use of two mission hospitals situated just over 20 km apart in Meru District, Kenya. Although it had been expected that people would travel to the hospitals from further away, in other words that their catchment areas would increase, this hypothesis was only partially supported by the evidence. The main effect of the improved road was to afford better and quicker access from within the existing, and partially overlapping, catchments. Although the catchments for in-patients did increase, the average distance travelled to each hospital barely changed after the road improvements had been completed. The proportion of patients using each hospital declined markedly with increasing distance, but this 'distance decay effect' was also not affected by the road schemes. Shortening of one road had a more marked effect than reduced travel costs, which in any case formed a small proportion of total treatment costs. However, transport cost reductions leading to greater hospital utilisation had been anticipated as a major benefit in the road improvement appraisals.

Far more important determinants of utilisation patterns proved to be economic and institutional factors, especially the cost of hospital treatment for in-patients. Religious affiliation was also a significant deterrent to attendance by non-Christians or Christians of other denominations, despite the fact that both hospitals operated non-discriminatory policies. Income levels and social structure must therefore be taken into account by health and transport planners. Health impacts of a very different kind are examined in Case study E.

Case study E

Highway construction and health in the Papua New Guinea highlands

The Southern Highlands Rural Development Programme in Papua New Guinea was a geographically defined integrated rural development programme designed to expand and diversify the economy of this poor province in west central PNG and to link the area with the country's economic core. Funded by the World Bank, it comprised the construction of a highway and feeder roads, the establishment and rehabilitation of coffee and tea plantations and associated factories, agricultural research, schools and hospitals.

Construction of the Mendi–Koraba highway was advocated in terms of precisely the mix of objectives discussed in the main text above, namely enabling the government to meet basic needs, tackle poverty and malnutrition, provide basic services, promote economic development and to enable political consolidation. Economically, it was justified on traditional cost–benefit analysis criteria, showing an 8 per cent economic rate of return at initial appraisal and an astonishing 39.1 per cent when the project was completed in 1981. However, the factors included in the analyses concentrated on savings to existing air travellers on the route, benefits to new travellers and to freight. Other impacts were mentioned but not included, probably on the grounds that they are too difficult to measure, are perceived subjectively and/or take a long time to manifest themselves.

To be sure, over 100,000 people were linked to the national economy, coffee production rose markedly over the first few years after completion of the highway, and the supply of goods and services was greatly facilitated. However, the fallacy of the very narrow traditional view of costs and benefits was clearly illustrated by the dramatic and rapid emergence of important unforeseen impacts within a few months of construction commencing. In particular, forest clearance along the highway route allowed malarial mosquitoes to penetrate into parts of the highlands where they had not previously occurred. They then bred rapidly in borrow pits, wheel ruts and other water-filled depressions. Infected workers from outside the area also contributed

Case study E *(continued)*

somewhat to the spread of the disease, as did higher mobility among previously isolated communities. In the district most directly affected, reported cases of malaria rose dramatically during 1978 and early 1979, culminating in over 70 deaths in the population of about 1,000 in the Hwim area. The proportion of blood slides taken which were found to have malarial parasite infection rose from 16.5 per cent in January 1978 to 71.9 per cent that December. By January 1979, the vulnerability and high infection rate among young children was appreciated, and the subsequent deaths triggered a swift anti-malarial spraying programme (using highly toxic DDT) by the provincial authorities; this succeeded in bringing the outbreak under control.

The large numbers of non-local construction workers and the increased local mobility led to rapid increases in the reported levels of sexually transmitted diseases (STDs), especially gonorrhoea and syphilis in the Southern Highlands. National incidences of these diseases had been increasingly on a steady and steep trend (Figure E.1), but the sudden surge in Southern Highlands Province is dramatic and can have important effects, especially for the health of women. Gonorrhoea has no symptoms in women until it reaches an advanced stage, by when sterility may already have resulted. Nationally, over half the cases reported in 1984 were in the 15–24 age group, and 78 per cent were male, among whom the symptoms become evident earlier on and who are often more mobile than women, and thus able to seek treatment relatively anonymously away from home.

Another impact of the highway was an upsurge in 'tribal' fighting on the plateau, resulting in at least thirty-five deaths and much destruction of crops, livestock and homes. Longstanding rivalries appear to have been aggravated at least partly by perceptions of malaria and STDs having been introduced by outsiders and their sorcerers.

The lessons of this case study which, while perhaps dramatic, is by no means unique, are that new road construction needs to consider a far wider range of positive and negative developmental impacts, and that these must be incorporated and given due weight

Case study E *(continued)*

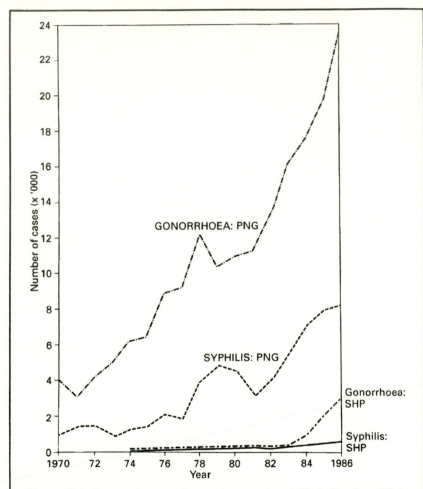

Figure E.1 Reported cases of gonorrhoea and syphilis in Southern Highlands Province (SHP) and all of Papua New Guinea (PNG), 1970–86
Source: After Crittenden and Lea (1988), Figure 2

within the relevant appraisal and decision-making processes. By no means all impacts are foreseen, and many direct and indirect consequences occur far more rapidly than economists would have us believe.

Case study E *(continued)*

Source: Crittenden, R. and Lea, D.A.M. (1988) 'Roads, cash cropping and communicable diseases: coffee and tea, malaria and STDs in the Southern Highlands Province of Papua New Guinea', paper presented at the International Geographical Union Congress, Sydney, August.

Rural feeder roads

It should be noted at the outset that not all so-called feeder roads actually penetrate previously inaccessible rural areas, as the name implies. It is quite common for the term to be used merely to describe low-cost, low-volume laterite or gravel roads in rural areas. The impact of rural feeder roads (however defined) on accessibility and various aspects of rural development was well researched by a number of studies undertaken mainly in Africa during the early 1980s, by both transport economists and geographers. Accessibility was defined in various specific ways for data-gathering purposes. For example, one group of researchers in Ghana used the cost of moving one headload of produce from each village in the study area to Kumasi, the Ashanti regional capital. This was the form of transport used from the least accessible villages, so where vehicles were used for some or all of the journey, the data were converted to enable direct comparison. Other studies have used mean travel time to the regional or subregional centre, measured using a specific form of transport, or have imputed such times by dividing the actual (or approximate) distance by the average speed of a particular form of transport.

Research methodologies have also varied, the majority using some form of correlation and multiple regression analysis to test the relationship between changes in accessibility and changes in other variables such as trip rates (the number of trips per unit time), the area under cultivation with particular crops or using recent technological innovations, agricultural output, agricultural and other locally available product prices, vehicle ownership and household incomes. This kind of methodology has its limitations, being dependent on many estimated values (whether by respondents or researchers), omitting many wider social and politico-economic effects, and producing results which frequently lead researchers unjustifiably to infer causation from statistical association. A few studies have deployed intensive field interviews in more qualitative ways, seeking to draw on life histories, diary keeping or other ethnographic methods, with a blend of quantitative and more qualitative, experiential data.

The results have been mixed, and require care in interpretation against the specific local context, the research methodology used and questions addressed. One of the central challenges to such research is that most rural farmers nowadays market at least some produce if conditions are at all conducive. If their landholdings are large enough to permit the growing of some surplus, the response to changing market conditions and prices can be quite flexible. In many areas, there is thus a spectrum of farming practices in terms of the proportion of produce grown for own consumption or sale, the size of landholdings and the extent of essentially traditional multicropping and intercropping systems as opposed to more 'modern', commercially oriented monocropping systems.

This makes direct correlations and comparisons difficult. For example, use of commercial fertilisers, herbicides and pesticides, as well as new higher yielding varieties of seeds, is far easier and more appropriate in commercial agriculture at almost any scale than in mixed cropping systems. So any attempt to measure the extent of use of such inputs as an indicator of agricultural development and to relate this to differences in accessibility, in terms of contact with agricultural extension services and commercial or co-operative sales and purchasing depots, has to take the variations in farming system, land tenure, local soil and environmental conditions and other relevant factors into account.

Available research underlines the importance of the wider social and political economy, and also the specific innovation being examined. The Ghanaian study referred to earlier, undertaken in a fertile, densely populated region with strong indigenous and commercial traditions and relatively well developed infrastructure, found that:

Within the range of accessibility considered, if anything, the least accessible villages appear to be more agriculturally developed than the most accessible villages. The least accessible villages had larger farms, grew more cocoa [a key cash crop] and sold a greater proportion of the crops they produced. They also devoted more labour to farming (per member of household). . . . However, the overall strength of the relationships found was weak. No evidence was found to suggest that the less accessible villages suffered any disadvantages in obtaining insecticide, fertiliser, using tractors or gaining extension advice. However, poor accessibility might adversely affect agriculture in an important way, through the inability to obtain finance [although applications for financial assistance were strongly age-related].

(Hine, Riverson and Kwakye 1983: 23)

The key to such apparently paradoxical results lies in the broader economy:

Villages with better accessibility appear to be more dependent on non-agricultural activities for their livelihood. The development of . . . rural industry and more particularly the provision of rural services are, at first sight, more likely to be dependent on good accessibility. . . . The study supports the conclusion that where a road investment induces only a relatively small change in transport costs and market prices (such as would arise, for example, from the upgrading of an existing track or earth road) then correspondingly little impact on agricultural development may be expected.

(ibid.: 23)

Obviously, where substantial journey distance (or time) savings occur, or motor vehicles can replace human headloading or draught animals, the relative change in transport costs will be greater and the likely impact therefore more marked. In terms of 'capturing the peasantry' or stimulating rural development, it thus seems that, while feeder roads may prove effective in the absence of existing roads that are passable to vehicles, the upgrading of existing roads is less effective. Of course, upgrading may occur for other reasons, including political pressure and patronage, and as part of wider rural development programmes. In the case of such integrated or other multifaceted development programmes, the overall impacts may be substantial but it becomes difficult to isolate the effect of each component, including the road construction or upgrading element.

Moreover, such schemes may be undertaken in areas of high population density, political significance or existing high development potential rather than those of lower potential or great absolute need, in other words on a 'best first' rather than 'worst first' basis. Foreign donors are often keen on this as it enhances the chances of success and the likely magnitude of measured change (Case study F).

Case study F

Feeder roads and integrated rural development in eastern Sierra Leone

As Airey (1985) found in Sierra Leone, significant elements of externally funded and led integrated rural development programmes may prove unsustainable in the long term in poor countries, especially if they are not locally appropriate. By contrast, spontaneous changes induced by the feeder road programme, e.g. the diffusion of Amazon cocoa, pit latrines and corrugated zinc sheets, were more successful and sustainable. Even the least accessible communities adopted these and could be classed as 'modernising communities'. The wider use of pit latrines and more durable building materials like zinc sheet roofing or cement veneer on walls was having a noticeable impact in terms of improved health. Use of such materials is often taken to reflect rising incomes (or to reflect income distributions), but the study showed that the agricultural activity and incomes of the localities with the highest adoption rates of cement were no higher than elsewhere. In this particular case, the adoption of cement was highest adjacent to the new roads, and declined with increasing distance from them. This reflected the perishability of cement in the humid tropical environment where appropriate storage was not available. By contrast, zinc sheeting, which also had to be purchased, showed no such 'distance decay' away from roads, being portable and non-perishable.

Another significant finding was that most households valued the roads for social as much as economic purposes and, on average, made relatively little use of them even five years after construction. Furthermore, the absence of a maintenance element in the programme was leading to rapid erosion of the feeder system, which would become unusable within a few years unless the situation could be rectified. Few of the roads constructed within the programmes studied were true feeder roads as defined above. Airey ultimately felt that low-cost tracks would have been more economical and at the same time more locally appropriate to the current state of development and needs in the area.

Case study F *(continued)*

Source: Airey, A. (1985) 'The role of feeder roads in promoting rural change in eastern Sierra Leone', *Tijdschrift voor Economische en Sociale Geografie* 76(3): 192–201.

The final issue to be covered in this section relates to the actual technology and method used in road building. Particularly with local and so-called feeder roads where traffic volumes are relatively low and technical standards not necessarily so exacting, opportunities exist to increase their developmental impact and employment-generating role by using labour-intensive construction methods rather than conventional labour-replacing machinery. The idea, which has been implemented in numerous countries over the last decade or so, often with encouragement from the International Labour Organisation (ILO), is to provide local

Plate 4.3 The Onaanda labour-based road construction experiment, northern Namibia, 1992

Plate 4.4 Conventional road construction methods, Omusati Region, northern Namibia

employment and on-the-job training in areas of high un- and under-employment, and ensuring that a far larger proportion of expenditure on the roads accrues locally and enters the local economy rather than being paid to large contractors (Plates 4.3 and 4.4). In addition, a local capacity for subsequent road maintenance is built up, with relevant individuals being allocated responsibilities for short stretches of the road near their homes.

Like many apparently 'radical' departures from conventional practice, this innovation has met with resistance from vested interests in the contracting sector, as well as from traditional transport planners and engineers and some politicians. Their opposition has centred on the supposedly inferior product and the longer construction time. In practice, results have been mixed, with broad programmes often more successful than one-off experiments, as the necessary experience has to be built up and replicated. Labour management has proved the most difficult aspect, given the number of workers involved, their general lack of previous relevant experience and the project duration. Where single

Plate 4.5 An experimental twin track road surface of interlocking concrete slabs, Omusati Region, northern Namibia

project experiments have encountered problems and political or professional support has been lukewarm, the idea has commonly been abandoned, as is the case with many other forms of low-cost or low-maintenance road construction materials and methods (Plate 4.5).

Farm mobility and appropriate rural technology

The discussion so far has concentrated on off-farm accessibility and mobility, in other words between farms and local markets, service centres and larger towns. However, it is important also to consider the nature of on-farm mobility in the context of different landholding systems and patterns, and the implications for development of interventions at that scale. In densely populated, intensively farmed areas where the pattern of landholdings has undergone reorganisation and commercialisation, such as Singapore, Taiwan, parts of Java and Sumatra in Indonesia and more limited localities of other countries, mobility may not be problematic, especially if incomes are sufficiently high to enable widespread access to and use of motorised vehicles.

However, elsewhere it is common for farmers to have irregular patterns of landholdings, often with plots or grazing lands scattered over a considerable distance. This can arise for different reasons, depending on local conditions and land tenure systems. The principal reasons include:

- the fragmentation of original plots through inheritance necessitating the acquisition of additional land elsewhere to provide sufficient food;
- the need or ability to exploit various environmental niches by growing different crops at specific altitudes or distances from a river or other water source and thereby diversifying diets and spreading risk;
- exhaustion of soils or unsuitability of the terrain in some or all areas close to the homestead;
- undertaking a mixture of cropping and livestock rearing; or reflecting where access to land for expanding the scale of production could be obtained. This might be marginal land (in terms of slope, soil quality, access to water, etc.) cleared of vegetation for the purpose.

These reasons are both positive and negative. Hence efforts by development staff to regularise and consolidate landholdings on a comprehensive basis as part of a rural development programme designed to facilitate commercial production have often been opposed by the intended beneficiaries if they perceived such changes as likely to undermine deliberate survival strategies, to require the loss of valued assets such as fruit trees (which take years to replace) in particular localities, to increase the level of risk or displace a significant proportion of villagers. This indicates again the complexity of development issues and the need for sensitivity to local conditions, practices and preferences. Small or moderately sized projects and programmes are generally more flexible in design, able to be made locally appropriate, to involve local participants substantially in design and control, and hence to have a greater prospect of success.

Let us therefore examine evidence from farm surveys in various parts of Africa and then discuss the implications of possible interventions. It is common for 10 per cent or more of the average working day to be spent walking between home and fields in various parts of the continent. One study in Calabar, Nigeria, put the figure at no less than 30 per cent. Table 4.1 gives more detailed information. Average distances between homes and fields in four districts of northern Nigeria were found to

Table 4.1 Distance and travel times to fields from homesteads, northern Nigeria and Geita, Tanzania

(a) Northern Nigeria

District/state	Average distance (miles)	Walking time (hours/year)
Mokwa	3.8	600
Kutigi	2.4	350
Katsina	1.0	100
Zaria	< 1.0	100

(b) Geita

	<15	15–29	Walking time (minutes) 30–44	45–59	60–89	90–120	Total
n	81	99	40	76	17	10	323
%	25	31	12	24	5	3	100

Source: Various farm management surveys summarised in McCall, (1985) 'Accessibility and mobility in peasant agriculture in tropical Africa', paper presented at Institute of British Geographers' Annual Conference, University of Leeds, January

range from under 1 mile to almost 4 miles. However, more important is the relationship between distance and the time taken to cover it. The nature of the terrain is obviously important in this context, but the data suggest that total annual walking time increases more than in proportion to increased distance; in other words, there is a geometric relationship. This is hardly unexpected, as people tend to walk more slowly over longer distances, to tire (especially if carrying significant loads) and possibly to stop along the way. The data from a survey of five villages in Geita, Tanzania (Table 4.1), reveal that 56 per cent of all the fields were within a half-hour's walking time of the villagers' homesteads but that 44 per cent required more than that, with 8 per cent over an hour away. The implications of such data are profound.

1 There is a direct trade-off between time spent in transit and time and energy available for cultivation and related tasks or other activities. In many rural contexts, the working day is still defined by the length of daylight, which provides a firm limit on outdoor work even if domestic lighting is available.

2 Such average data hide a pronounced implicit gender bias. Given that women in most African societies are responsible for the bulk of cultivation tasks apart from heavy work like bush clearance and ploughing,

they bear a disproportionate share of this burden. They also have primary responsibility for childcare, domestic work, collection of fuel and water, and care of the elderly, i.e. social reproduction. If time spent on cultivation and associated transit increases or is excessive, the health and nutritional status of the household could well suffer.

3   The intensity and regularity of cultivation, crop protection and related activities have been shown to decline markedly with distance from the homestead. This means that the quantity and value of crops grown at a distance are likely to be lower than of those grown nearby. Nevertheless, they may form an important element of the total household food budget or marketable output. Clearly, the more reliant a household is on distant fields, the more vulnerable it is likely to be.

Improving on-farm mobility

In situations like these, even minor innovations that reduce the time taken to move between home and fields or to collect water and fuelwood can have a marked impact on the pattern of daily activities and consequently upon quality of life. If access to water is one problem, then construction of a well or connection to another safe source of supply within the village can increase the quantity, quality and reliability of supply, avoid the health problems associated with contaminated water, and increase the time that the women – who again are usually responsible for water collection – can devote to other tasks.

Short of redesigning the field system, however, the locations of a household's plots are often fixed. Tractors, pickup trucks or other conventional four-wheeled motor vehicles are expensive to buy and maintain, not alway suitable for the terrain, may be of little use in the absence of passable roads, and are almost certainly beyond the financial means of most poor rural households. One option is to form a collective or co-operative and to purchase such a vehicle to be shared among the villagers or members. This has been done in many regions around the world, but requires good organisation, goodwill and maintenance capacity. Regular supplies of fuel may also be problematic in inaccessible areas.

Other approaches, likely to be cheaper and more appropriate, especially in remoter areas or uneven terrain, are, first, to examine the footpath network for possible ways of shortening or facilitating the walks required. This may have greater potential where land tenure is at least partly communal, in that permissions may be easier to obtain and rights of transit more readily maintained. Second, appropriate transport aids and vehicles can be investigated. They have the potential to reduce travel time and/or to increase the loads that can be carried and the distances over which they can be transported. They fall into three broad categories, namely draught animals, vehicles which are powered by humans or animals and those powered by inanimate energy. Each will be discussed in turn.

First, however, we should return briefly to the subject of appropriate technology and technological imperialism, which was addressed in Chapter 3. Clearly, this is a potentially emotive issue. Just as the modernist obsession with importing the latest hi-tech transport solutions has sometimes been condemned as imperialist in contexts where they are inappropriate, i.e. cannot be afforded, maintained or are otherwise incompatible, attempts to promote low-cost, low-technology solutions have occasionally been criticised as seeking to perpetuate poverty and 'second-best' conditions in Third World countries.

If this were done across the board, or all newer, higher technology solutions were rejected out of hand, this might be a fair concern. However, the central point about appropriate technology is that *the most appropriate* technology and solution for the circumstances should be adopted. This does not rule out any specific technology as such. However, in many situations, simpler, less sophisticated and cheaper solutions, but which nevertheless represent an advance over what, if anything, has been used there previously, are often the most appropriate. This is sometimes known as intermediate technology. Indeed, there is a special British-based charitable development organisation dedicated to the promotion of such technological choices appropriate for developing countries (and also for Britain). Known as the Intermediate Technology Development Group (ITDG), it conducts research, gathers and disseminates information around the world.

Draught animals

Draught power is often essential for heavy farm work, such as ploughing, pulling carts or pack carrying. Some animals can also be ridden directly. However, many of the poorest rural peasants and smallholders lack access to them, being unable to afford their upkeep or to replace those lost through disease or drought. Replacing such lost draught animals is now recognised as an important component of post-drought

Table 4.2 Transport characteristics of selected draught animals

	Size (kg)	Speed (kph)	Average load (kg)	Max load (kg)
Mule – light	270	7.2	34	50
– heavy	650	7.2	82	115
Donkey	200	5.0	50	135
Dromedary – light	350	4.0	96	140
– heavy	600	4.0	165	240
Horse – light	350	5.6	40	55
– heavy	640	5.6	75	95
Ox – light	250	3.5	30	65
– heavy	700	3.5	90	175

Source: McCall, (1985)

recovery programmes in arid and semi-arid zones, where vulnerability is great. However, the same rationale could be applied to poor households who have not previously had any animals. The most appropriate animals for a particular locality depend on the environment, prevalent stock diseases and the nature of the work for which they will mainly be used. If human mobility without any supplementary vehicle is of prime concern, then a horse, mule, dromedary or camel is suitable, whereas for carrying small loads or pulling carts, a horse, mule or donkey might be appropriate. For heavy work like ploughing large fields and pulling wagons, oxen or water buffalo are probably best.

Animals are generally quite versatile and adaptable, can carry surprisingly heavy loads, as indicated in Table 4.2, and do not require special paths or roads. They do, however, require daily upkeep in the form of food, grazing land and/or a safe pen or stockade. Periodic veterinary attention is also important. Although the various animals listed in Table 4.2 vary considerably in their average speed, this is a relatively unimportant factor except for long-distance riding or pulling. Even so, stamina and endurance are likely to be more significant factors.

Unpowered vehicles

The range of such vehicles is extremely wide, but can be categorised into those, like conventional carts, trailers and so forth, designed to be pulled by animals, and those where humans provide the locomotion, such as bicycles, tricycles and handcarts.

The range of wagons, carts and trailers designed for pulling by animals is vast, showing diversity of construction materials, engineering skills and sophistication, and adaptation to local conditions. One of the most widespread vehicles for traction by horses, mules or donkeys is the so-called Scotch or donkey cart (Plate 4.6). The basic structure is conventionally wooden, with an old car or truck axle underneath and some form of wooden bench across the top. These carts are used for personal and freight transport over considerable distances. Design improvements and innovation are often evident, as in the increasingly common recent phenomenon observed in Namibia of replacing the wooden structure with the rear half of the chassis of a wrecked pickup truck (Plate 4.7).

Handcarts also come in numerous designs, and are generally best suited to relatively smooth tracks and flat ground or very low gradients, such as in and around villages rather than on farms or over long distances. On-farm equivalents would be wheelbarrows and adaptations of that basic design which offer greater or differently shaped load capacity and better stability.

Bicycles represent one of the most widespread and versatile forms of rural transport, greatly increasing mobility and accessibility both on and

Plate 4.6 A traditional Scotch cart, Khomas Region, Namibia

Plate 4.7 A new Scotch cart made from the rear of a pickup truck chassis, Otjikoto Region, northern Namibia

off roads, enabling substantial loads to be carried as well, and requiring no fuel as such (Plate 4.8). However, they do need a significant initial investment for purchase – sometimes the equivalent of a month's income or more. At the same time, cheaper imports from China and Taiwan, in particular, have reduced prices in non-producer countries over recent years, while important second-hand markets recycle bikes at still lower prices. Different bicycle frame designs and tyre gauges exist, and are suited to different conditions, although mountain bicycles are a recent innovation and still too costly to have made an impression beyond the luxury market. No special facilities are required, although they should be ridden on paths or better tracks. Maintenance is easy and comparatively cheap unless tyre tubes have to be replaced frequently; vulnerability to punctures is the main drawback apart from the need for a degree of physical fitness to ride them. Numerous modifications have been made in different contexts to strengthen the frame for load carrying, to fit large panniers or other containers or even to attach small trailers behind the saddle.

Plate 4.8 Bicycles in Makutano rural market, West Pokot, Kenya

Tricycles and bicycles with a sidecar have similar features, except that the two parallel tyres at the rear restrict the terrain and type of track on which tricycles can be used. Their load-carrying capacity is, however, substantially greater, as trailers, load platforms or passenger compartments can be fitted. The range of designs is again wide; this form of vehicle is common in urban areas as well, and will be discussed further in Chapter 5.

Powered vehicles

This category includes well-known forms of transport like motor scooters and motorbikes, which are increasingly used in rural towns and villages. Newer designs like motocross bikes have broad tyres and are proving valuable for agricultural extension officers, wildlife conservation staff and others in rough rural terrain. In addition, there are many types of adaptations such as powered bicycles and tricycles, which offer the advantages of greater power on low-cost vehicles, with the possibility of pulling trailers and similar attachments to convey heavier loads or traverse greater distances than unpowered vehicles. The principal difference, however, is that petrol or diesel fuel is required, and this can be expensive for many rural households, as well as a greater initial outlay and maintenance requirement. Such forms of transport will also be discussed further in Chapter 5.

Conclusions

The relative importance of rural development is perhaps greatest in poor countries, which tend to have the highest proportions of their total populations living in rural areas, often in conditions of poverty. Although Africa is, on average, probably the most rapidly urbanising continent, this is occurring from a comparatively low base, since some 70–75 per cent of its 640 million inhabitants are still classified as rural. Only a handful of countries (Djibouti, Tunisia, Algeria, Libya and South Africa) have predominantly urban populations. In Asia the picture is very diverse, with the newly industrialising countries of Singapore and Hong Kong, Taiwan and South Korea now highly urbanised but poor countries like Nepal and Bhutan still overwhelmingly rural. Even India, the world's second most populous country (*c.* 800 million inhabitants), which boasts three megacities of over 8 million people and another of over 4 million, is only about one-quarter urban. Of the continents in the South, only South America has a minority of rural dwellers, about 40 per cent, although the proportion in the poorer countries like Bolivia, Paraguay and Guyana is higher.

Nevertheless, regional inequality (and the social and economic inequality which it represents) exists in virtually all countries of the South. The construction of trunk and feeder roads in 'remote' or poor regions has been seen by governments of all political persuasions and by donor organisations as an important tool for promoting development. Sometimes such projects are undertaken as one-offs but, especially with feeder roads, they have increasingly been provided as components of multifaceted rural development programmes. Labour-intensive construction techniques have sometimes been used in efforts to maximise the employment potential and hence the local developmental impact.

Many misconceptions about the relationship between new road construction and development persist and can be attributed to the continued dominance of modernisation theory and its offshoots in the transport field. Thus, for example, there need be no causal relationship between low accessibility or mobility and poverty. Equally, the mere provision of road connections does not guarantee that development (however broadly or narrowly conceived) will follow.

Research in a wide variety of countries and regions, encompassing trunk as well as feeder roads, has highlighted the complexities of these relationships and how dependent on local conditions the outcome is. Generally, the most significant improvements are recorded where no road previously existed rather than where poor roads or tracks have been upgraded; where transport costs comprise a substantial proportion of total costs of the relevant goods or services; or where transport improvements enable a significant time or cost saving. Even so, new roads are most often permissive rather than automatic triggers of development. Numerous factors determine whether innovations or new commodities become widely available or exhibit pronounced 'distance decay' away from access routes. The impact of new roads on local communities can also be partly or predominantly negative.

Mobility on farms or within communities is often neglected by transport and rural development planners, but the scope for minor, cheap and yet significant improvements is often substantial. Many peasants and smallholders spend considerable time and physical effort walking between their homesteads and fields; these can be reduced in a variety of ways. The wide range of draught animals, unpowered and motorised vehicles available was discussed, with relevant examples. Many of these are not specifically rural, being found in the same or similar form in cities of all sizes.

Key ideas

1 There is no necessary or direct causal relationship between transport improvements and development.
2 Inaccessibility and relative immobility are frequently associated with poverty but this need not necessarily be so and improvements can have very mixed impacts.
3 Transport improvements are not always implemented on the basis of identified need, but of political expediency, donor preferences and even the greatest likelihood of measurable success. Areas of high or moderate potential may thus be favoured over poor areas with low perceived potential, even if the relative impact in the latter would be great.
4 The extent to which a transport improvement or innovation promotes development depends very largely on a complex set of interrelation-

ships and local conditions. When new roads form one element of integrated rural development programmes, their impact is particularly difficult to isolate.

5 Generally positive impacts are greater where new roads are provided for the first time, where they make a substantial difference to the ease and duration of journeys, and where transport costs form a significant proportion of the total costs of goods and services.

6 A wide range of draught animals, unpowered and motorised vehicles is available, covering the full spectrum of technological sophistication, so that locally appropriate forms can be found for virtually any context, both on farms and for longer distance travel. Speed may be far less important than reliability, cheapness, lack of maintenance, the size or mass of load carried, and the terrain over which movement is required. Appropriate, intermediate technology generally offers a way forward rather than inhibiting development.

5
The urban transport challenge

Introduction

In this chapter the focus switches to the situation in towns and cities of the South. The range of transport modes and forms found in different contexts is outlined and the nature of the sometimes severe transport problems facing urban areas is discussed. Although there is in reality often more of a continuum from rural to urban areas than a neat division between them, the distinction is used here as a shorthand to highlight the more conspicuous contrasts. Small urban centres and villages in rural areas frequently have features typical of both urban and rural areas. This in no way negates the argument.

Equally, although towns and cities have some basic features in common, the diversity of urban forms, built and socio-cultural environments, economic bases and institutional contexts is great. Thus, for example, two cities of comparable population size and rates of growth may nevertheless face very different problems that should be addressed in different ways, as a reflection of their particular physical environments and conditions, local histories and lengths of existence, cultural and social compositions, politico-legal and institutional frameworks, positions within their respective national and continental urban systems, and so forth. Transport systems represent just one of these dimensions, and they too need to be understood in terms of the history, culture and functions of each city, not merely as a random selection of disparate technologies. It is arguably true that, despite elements of continuing diversity, transport systems and policies being pursued in large Third World cities are converging to some extent. This is not coincidental but the effect of increasing globalisation in production, consumption and communications, and the increasing role of international consultants, development agencies, Northern producers and national elites in promoting the adoption of specific transport technologies and traffic management approaches. Whether, and to what extent, they are appropriate in any given context is one of the issues to be explored in this chapter.

The importance of urban areas, and therefore the transport systems within them, derives from the increasing proportion of every country's population living there at least some of the time, the greatly increased proportion of economic value-added being produced within them, and the steadily increasing reliance of modern and modernising societies on mobility and transport. The hi-tech age is also increasing the role of cities as communications and transactional hubs, as links between domestic and international relations in many spheres. Although the widespread use of the computer, modem and fax is actually reducing the importance of a central urban location for many types of business and professional activity in the North, such technologies and the skills and resources to support them remain very geographically concentrated in cities in most of the South. There are, of course, a few notable exceptions, like the city states of Singapore, Hong Kong and possibly Taiwan, and the new 'extended metropolises' of Southeast Asia.

The populations of most sizeable towns and cities rely on a wide variety of transport, both public and private, motorised and non-motorised. The balance is not, however, static. Probably the most conspicuous trend since the Second World War has been the increasing reliance on motorised forms of transport – and over the last 20–30 years also in private motor vehicle ownership – in cities of the South. Although the latter process undoubtedly reflects rising disposable incomes, this is by no means the only important factor. These issues will be discussed below. Nevertheless, motorised public transport, both rail- and road-based, remains substantial almost everywhere. In megacities and some other large metropolises, public transport is actually experiencing something of a revival as a result of increasing traffic congestion, concerns over growing pollution levels and/or the needs of the large poor components of their populations. Various measures have been adopted to restrict private car traffic, to prioritise public road transport and to increase the speed, reliability and carrying capacity of the public transport system as a whole through the introduction or improvement of light rail and mass rapid transit systems on high volume routes. At the same time, non-motorised transport is experiencing divided fortunes. As discussed in Chapter 2, bicycling remains significant and is increasing in many cities in the South and North, although more traditional rickshaws and trishaws are under pressure from urban authorities in South and East Asia as part of their drive for modernisation and more rapid and supposedly efficient urban transport systems. These modes and the associated issues will be discussed in turn, although we start by considering the cheapest, least space-intensive and often most important form of movement over short distances, namely walking (Case study G).

Case study G

The importance of walking in Third World cities

The vital importance of walking, especially over comparatively short distances, in cities of all sizes is often overlooked. Indeed, many people do not even consider it to be a form of transport. Catering for pedestrian flows and their safety forms an increasing concern of urban transport planners today, although conditions and facilities are often poor or rudimentary in cities of the South. Crossing busy roads is generally far more dangerous than in cities of the North, where pedestrian crossings are better, and compliance with traffic regulations and law enforcement usually more widespread. Pedestrians also account for a far higher proportion of casualties in road-based personal injury accidents. Generally speaking, the poorer the city, the less adequate its pedestrian facilities are.

There have been surprisingly few systematic studies of the extent of walking; some data that do exist are summarised in Table G.1. From this, it is evident that something like 20–40 per cent of all trips in a range of tropical African capital cities plus Jos in Nigeria were made on foot in the 1980s, while in the four Indian cities the range was narrower, from 35–40 per cent. It is noteworthy that the percentages do not reflect city size in any systematic way; some of the largest cities recorded the highest level

Case study G *(continued)*

Table G.1 The importance of walking in selected Indian and African cities

	Population (million)		Proportion of all trips by walk	Average walking distance (km)
Jaipur	1.0	(1981)	39	1.2
Vadodara	0.7	(1981)	40	1.2
Patna	0.9	(1981)	36	1.3
Delhi	6.1	(1981)	40	1.1
Dar es Salaam	1.5	(1987)	25	1.7
Jos	0.4	(1986)	23	1.2
Douala	1.1	(1987)	28	1.2
Yaoundé	0.8	(1987)	30	1.7
Harare	1.3	(1987)	42	1.6

Source: Maunder, D. A. C. and Fouracre, P. R. (1989)

of walking. The physical size, density and accessibility of different areas within the city, and the availability and cost of public transport are likely to be far more important determinants. Furthermore, if the data in the table are generally representative, it appears that 40 per cent or so represents a ceiling for the proportion of all trips done on foot. While cycle ownership might also be thought relevant in this respect, it will be shown below that cycles are actually far more widely used in Indian cities than in the African cities under discussion here. Given that the mean number of trips per person per day across all these cities is comparable (between 1.5 and 2), the difference can probably be explained in terms of use of public transport and various private vehicles.

Naturally, walking trips are made predominantly over short distances, the average in all the cities in Table G.1 being between 1.2 and 1.7 km. This substantial uniformity reflects average human stamina and time considerations. The propensity to walk is also determined by trip purpose and the load, if any, that has to be carried. Moreover, it seems likely that low-income people make a higher proportion of their trips on foot than others. For example, data collected in the three intermediate Indian cities in Table G.1 suggested a figure of between 47 and 58 per cent, compared with about 40 per cent for middle-income earners and only from 23 to

Case study G *(continued)*

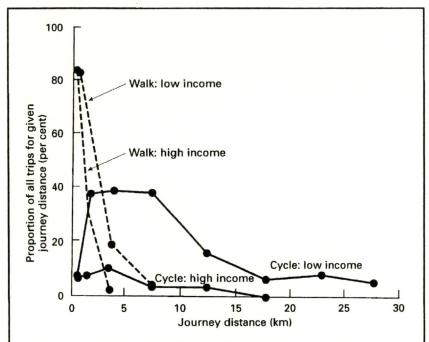

Figure G.1 Cycling and walking by distance and income group in Delhi
Source: After Maunder and Fouracre (1989), Figure 2

27 per cent in the high-income group. A similar relationship was found in Delhi (Figure G.1).

Source: Maunder, D. A. C. and Fouracre, P. R. (1989) 'Non-motorised travel in Third World cities', paper presented at the Institute of British Geographers' Annual Conference, Coventry, January.

Private vehicle ownership and use

As was pointed out in Case study A, car and motorcycle/moped ownership are not directly or linearly related to incomes; a few high-income countries have lower ownership levels than certain others which are considerably poorer. The variations in per capita incomes and vehicle availability within countries alluded to in Chapters 2 and 3 exist within individual cities too, although it is difficult to obtain data at this scale.

Most vehicle data are collected by licensing or registration authorities, which generally cover entire cities and perhaps even some surrounding peri-urban and rural areas. Conversely, if a metropolitan area has experienced rapid urban sprawl, the built-up area may now extend beyond the territorial limits of the city's administration. Besides, many vehicles are used or even based in places other than where they are formally registered. Short of carrying out extensive field surveys, there is therefore no direct source of data on ownership rates by social class, income or ethnic group or suburb. Even at the level of whole cities, data are seldom published.

Motor cars

Table 5.1 provides a sample of car availability in a range of cities, although the dates and reliability of the figures are unknown. Once again, the number of cars per thousand of the estimated urban population bears some relation to average urban and national incomes. However, the examples which contradict this assertion are almost as conspicuous as those which support it. The two Chinese cities have far lower car ownership levels than expected, as a result of the state

Table 5.1 Car ownership rates in selected cities

City	Private cars per 1,000 people
Shanghai	2
Beijing	9
Lagos	15
Bombay	20
Seoul	31
Jakarta	44
Nairobi	50
Bangkok	87
São Paulo	145
New York	218
London	318
Stuttgart	442

Note: Base years unspecified
Source: Armstrong-Wright (1993: 3)

socialist system prevailing there, although since the economic liberalisation of the early 1980s, the number of cars, especially in these two cities, has been rising steadily. Buses and bicycles still account for the vast majority of urban non-walking trips in China. Nevertheless, the major cities of China, especially Beijing, Shanghai, Guangzhou and Wuhan, are experiencing rapidly increasing urban traffic problems. Since few people can yet afford private cars, which cost approximately 18 times the average urban annual salary, most of the over 1 million now in use belong to the state, parastatal and joint venture corporations. The number of private cars rose from a mere 60 in 1983 to 50,000 a decade later. At the end of 1983, there were 8 million cars, lorries and buses in the country; by the turn of the century this figure is expected to top 20 million. Given that cycling has hitherto been so widespread, road systems are inadequate for such traffic volumes, and standards of driving are reportedly poor, leading to many accidents on the pedestrian-choked streets.

Seoul also has a considerably lower figure than the country's level of economic development might suggest (Table 5.1). This reflects urban and social factors, coupled with the efficient, reliable and affordable public transport system. Conversely, the figure for Nairobi is substantially higher than that for Seoul, despite Kenya being a low-income country (ranked 19th poorest on GNP per capita by the World Bank's *World Development Report 1994*) compared with the Republic of Korea's NIC status. In this case, colonial history and more recent urban sprawl, coupled with a poor public transport service traditionally catering very largely for low- and lower middle-income residents, hold the key.

Similarly, the data on the three Northern cities illustrate the importance of factors other than average incomes. Despite its urban layout, which is well designed for motor cars, and historically wealthy image, car ownership in New York is less than half that in Stuttgart, on account of topography, congestion and restricted parking, a traditionally good public transport system and a large low-income population.

Such aggregate data conceal some potentially very significant intra-urban differences, not only by geographical area, but by class, ethnicity and gender. On the basis of informal observation, men appear almost universally to have greater access to cars – especially as drivers – than women. This applies across countries, cities and social categories, as a reflection of traditional gender divisions of labour, and the social and religious customs which commonly reinforce or determine them. Amongst middle- and upper-income groups in some advanced industrialised countries, this is now changing, but men are almost invariably still the registered owners in most cases, even if women make more use of cars. An extreme example of gender bias in religious and cultural constraints is provided by Saudi Arabia, where women are prohibited from driving cars at all.

Motorised two-wheelers

Motorcycles, mopeds and scooters offer far cheaper and less congestion-inducing forms of private passenger transport, although they are designed for one or two people at most and can carry only limited freight loads without some trailer or other adaptation. The national ownership data in Table B.1 show that, as with car availability, there is no simple or direct relationship with incomes. As a generalisation, it is probably fair to say that motorcycle and moped use tends to increase significantly as average urban incomes rise in lower middle- and some upper middle-income countries. However, this trend is likely to be reversed at a certain point when cars become more affordable to a wider section of the population, as these are perceived to be superior vehicles and conspicuous status symbols. Again, though, this is too simplistic, for ownership levels are rising in response to the artificially high cost of cars and vehicle licences in heavily congested metropolises like Singapore, despite the continued increase in average incomes (Plate 5.1). Use of such forms of transport is also more subject to social and cultural influences than cars, especially in Muslim and certain other 'traditional' societies. In particular, there may be sanctions against women riding them. Unfortunately, no city level ownership data are available.

Non-motorised transport

Bicycles

Especially among low-income groups, non-motorised private transport remains important. As in many rural areas (see Chapters 2 and 4), bicycles are widely used, especially in India, China and parts of Southeast Asia, on account of their ease of maintenance and flexibility (Plate 5.2). The relative cost of purchasing a bicycle is obviously important

Plate 5.1 Rapidly rising car and motorbike ownership levels pose severe congestion and pollution problems in the city state of Singapore

Plate 5.2 Sacks of raw cotton being transported by bicycle in central Bombay. Handcarts are also being pushed across this busy intersection

too: in India one costs approximately 60 per cent of the average owner's monthly personal income, whereas in West African cities the figure ranges from 1 to 3 months' income. This reflects various factors, not least the need to import bikes in Africa. Ownership levels in Chinese cities averaged 460 per thousand people, while in India rates of up to 200 per thousand people were recorded. By contrast, in the mid-1980s, ownership levels in several African cities (Jos, Yaoundé, Douala, Dar es Salaam and Harare) averaged only 20 per thousand. Cycle ownership appears to be growing rapidly: data for Ghana were given in Chapter 4, while in China their numbers increased by some 6–7 per cent per annum between the mid-1970s and mid-1980s, and in Delhi by 4 per cent annually during the 1970s.

The usefulness of bicycles is maximised for journeys of up to a few kilometres, although many cyclists plainly travel much further. A study in Delhi in the early 1980s found that the average work-related cycle trip was 7.6 km, and that almost 40 per cent of work trips in the distance

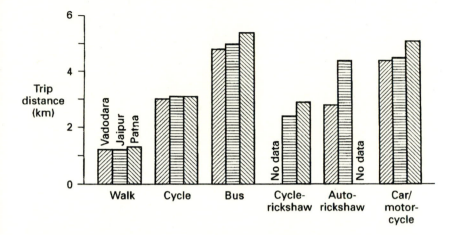

Figure 5.1 Trip distance by mode in three intermediate Indian cities
Source: Based on data in Fouracre and Maunder (1986)

range of 2–8 km by all income groups were by bicycle (Figure G.1). Among high-income earners, the figure was only 10 per cent, however. As Figure 5.1 shows, the average cycle distance in the three intermediate Indian cities was just over 3 km. Another study of fourteen intermediate Indian cities found that between 10 and 30 per cent of all trips were made by bicycle; the highest figures recorded were 34 per cent in Lucknow and 32 per cent in Jaipur. Congestion, pollution and safety risks often become deterrents to widespread use, especially as cities grow and car ownership rises. In the absence of dedicated cycle lanes or other facilities, cyclists often account for a disproportionate share of road accident injuries and fatalities.

Draught animals

Draught animals, especially horses, donkeys and mules, are still used in some areas for personal transport or for pulling Scotch carts, trailers and other forms of wagon in the course of work; bullocks are the principal draught animal in South Asia (Plates 4.6 and 5.3). In addition to income, ethnic and gender differences in use are strongly evident, as is caste in India. In addition, numerous types of handcart are used by poor people to move goods in cities of all sizes; this is very arduous labour,

Plate 5.3 A bullock cart in Bombay

Plate 5.4 Boxes of fresh produce being taken to a pavement market in Byculla, Bombay, by handcart

Plate 5.5 This design of handcart, here being used to transport crates of empty beer bottles in central Nairobi, is found in many African countries. It is widely known as the 'Mali cart'

especially in tropical conditions (Plates 5.2, 5.4 and 5.5). Again, unfortunately, no reliable data exist on the numbers or other characteristics of such forms of transport in individual cities.

Carts and wagons

The issues of appropriate and intermediate technology discussed in Chapter 4 also apply to urban areas, although the intermingling of diverse forms of motorised and non-motorised transport on congested urban streets is often frowned upon by local authorities and transport planners. They tend to regard it as problematic on account of the various vehicle types' very different characteristics, speeds and road requirements. In particular, human- and animal-powered vehicles are blamed for causing congestion because of their supposedly slow speeds and less predictable behaviour. Up to a point this may be true; some form of segregation may be appropriate. Nevertheless, the mean traffic speed in

Plate 5.6 The wide range of motorised and non-motorised traffic on Bombay's streets is unusual. Here lorries, a taxi and a scooter try to pass a cyclist, a labourer pushing a handcart, a bullock pulling a tanker trailer and a specially designed tricylce for a legless invalid

central Bombay, where no segregation or other restrictions are enforced (Plate 5.6), is no slower than in congested central London (*c*.11 miles per hour), with its range of traffic management measures.

However, many South and Southeast Asian governments, in particular, are endeavouring to phase out or ban non-motorised carts, trishaws and tricycles as unworthy of modern societies and the source of much congestion (see Case study G). This is an intriguing proposition, reflecting modernist thinking and planning paradigms, in terms of which the traditional and indigenous forms of transport, that (apart from dung) are non-polluting and do not rely on fossil fuel, are seen as inappropriate and alien intrusions into the home of the internal combustion engine. Such extreme policies seem inappropriate, especially as awareness of the problems inherent in emulating Northern urban transport systems uncritically is growing. More creative and flexible approaches are called for, in an effort to find locally appropriate solutions which incorporate roles for diverse forms and modes of transport. This is one of the most pressing urban transport challenges currently facing urban planners and managers in the South.

However, there is little evidence to date of a tangible shift away from conventional approaches. The dominant current policy trend is to deregulate and privatise public transport in line with international fashion and pressure, in the hope or expectation that the supposedly more efficient system that emerges will overcome the shortcomings of public sector operations. These policy issues will be discussed in Chapter 7.

Public transport

The wide array of public transport services can be categorised on the basis of mode, the technology used and the type of service provided. Hence the most basic distinction is between road- and rail-based transport. Each of these subdivides in turn. Under the former we will examine taxi and taxi-like, bus and bus-like forms. Under the latter, consideration will be given to conventional trains, metros (sometimes called mass rapid transits or MRTs) and light rail transits (LRTs).

While in cities of the North the distinction between taxis and buses is usually clear-cut, both visually and in terms of service characteristics, in many cities of the South there is considerable overlap between these categories. Indeed, it may be more useful to think of continua in two dimensions (technology and type of service) as represented in Figure 5.2. The extremes – which correspond to conventional licensed and regulated taxi and bus services around the world – are easy to identify. Essentially, a conventional taxi/minicab offers a car for individual hire at predetermined rates (either fixed or metered), with the route determined by the passenger's destination. By contrast, a stage bus provides a regular, scheduled service on fixed routes, usually with regular, predetermined stops, at predetermined flat fare or distance-based rates. It is used by different passengers up to the vehicle's capacity – although overloading is common in poor countries.

However, between these ends of the spectrum are numerous intermediate forms, sharing some characteristics of both. These straddle the two axes in Figure 5.2. For example, in many cities shared taxis using either fixed or variable routes account for a considerable proportion of public transport journeys. Equally, minibus services may operate on variable

	Bus-like	**Taxi-like**
Conventional	Single-decker bus Double-decker bus Articulated bus Minibus	Sedan/Saloon car ('Minicab')
Unconventional (paratransit)	Minibus *Jeepney* *Bemo* *Becak* *Tempo*	Hand rickshaw Cycle rickshaw (*cyclo,* trishaw) Autorickshaw (*tuk-tuk*) Horse tonga/*calera*

Figure 5.2 A matrix of road-based public transport modes

routes or a system of request stops, and use different fare systems. Many intermediate forms of public transport use unconventional vehicles and technologies, ranging from minibuses, adapted jeeps in Manila (the famous *jeepneys*) and converted pickup trucks (such as Kenya's rural *matatus*) to rickshaws and cycle rickshaws (like the *cyclos* of Phnom Penh and *becaks* of Surabaya). These unconventional forms are also known as 'paratransit', and often form extremely important elements of urban (and some rural) public transport systems.

Buses

These have long formed the most important single mode of road-based public transport, as they do not involve sophisticated technologies, can accommodate substantial numbers of people, and operate at modest fares. Capacity is extremely important in contexts where a high proportion of the population lack access to private transport and where urban railways are either non-existent or restricted to a few routes. Well-organised public transport systems involve interchanges between rail-based and bus services to facilitate feeder services and to ensure adequate coverage of the urban area. The way in which the stations of Singapore's MRT (Mass Rapid Transit) have been designed as bus interchanges provides a good example of this in a densely populated metropolis where average incomes are now high.

The range of bus designs and capacities has increased over the last decade or two, partly as a response to changing cost structures and in order to provide more flexible services in increasingly congested cities. In particular, minibuses seating between 12 and 25 passengers are now quite common. These may be operated as independent services by separate companies, or as part of a larger, more diverse bus fleet. Minibuses are particularly suited to the operation of low-volume routes, off-peak and feeder services, and to serving areas like indigenous walled towns, common in the Middle East, South Asia and West Africa, or unplanned or upgraded shantytowns, where the roads are too narrow, winding or uneven for conventional buses. They are also generally faster and spend less time at bus-stops. This rising popularity of minibuses is not restricted to the Third World, being widespread in Europe too.

Conventional single- and double-decker buses are often used within the same fleets, although double-deckers are absent from many cities like Nairobi, where their greater passenger capacity would be useful. Single-deckers are more widespread and more versatile, being less restricted in terms of height clearance under bridges and more stable on steep gradients. Standard single-deckers generally have a capacity of roughly 40 seated and 20-30 standing passengers, compared with perhaps 70–80 seated and 30–40 standing in double-deckers, giving totals of 60–70 and 120 respectively. Large double-deckers may hold up to 170 people. Articulated single-deckers and a variety of locally designed articulated or rigid tractor-trailer buses are used in some cities. These can carry similar numbers to double-decker buses. At peak times, however, overloading by up to 20–40 per cent is common, and it is not unusual to find double the number of legally permitted passengers in some cities.

Bus traffic is heaviest on major urban arterial routes, which are also particularly prone to rush-hour congestion. In some metropolises, like Bangkok, Bombay, Cairo, Jakarta, Manila and Mexico City, serious congestion has become the norm for much of the day. This greatly reduces the efficiency of all road-based modes, but may make buses less attractive options. However, where traffic management measures such as bus lanes (operated either at peak times only or all day), segregated busways or other forms of priority have been introduced, the average speeds and relative attractiveness of buses are increased. Under such circumstances, well over 20,000 passengers can be carried on a road per hour. Hourly flows of up to 15,000 and 20,000 passengers are common on roads with one and two lanes of mixed traffic in each direction respectively (Armstrong-Wright 1993: 7).

Traditionally, urban bus services have been run predominantly by the public sector: urban local authorities, public utility corporations or parastatals. Sometimes they have held an effective monopoly and in other places there has been competition between public and private operators, although frequently not on even terms as a result of subsidies or other advantages. There are some cities, such as Buenos Aires, where the bus service has been in exclusively private hands for many years. Approximately 15,000 locally manufactured 21-seater buses with a total capacity of 60 people, provide over 80 per cent of the estimated 10 million daily public transport trips in the city. Each vehicle carries a daily average of 850–1,100 passengers. The operation is divided into 300 route associations (*empresas*), which control schedules and the rotation of drivers around the routes to equalise revenues among them. The service is profitable and service quality reportedly good (Armstrong-Wright 1993: 14).

Changing bus policy

Several studies have shown that private bus operations (both urban and rural) provide generally more efficient and sometimes cheaper services, with fewer vehicles off the road at any one time, higher passenger load factors, quicker turnarounds and less overstaffing. In practice, where conditions are comparable, the difference in performance under differing extents of regulation, and between public and private operators, is often variable. As Table 5.2 reveals, the verdict is also strongly dependent on which particular variable(s) are being compared. Perhaps the most directly comparable case is between the public and private operators in Dar es Salaam, where passengers per bus and per km operated, and distance operated per bus per day are very similar but the differences in percentage vehicle availability, staff per vehicle, average passenger lead time (stopping time per passenger), average load factor and the cost structures are substantial.

Most public sector bus services have increasingly been plagued by some combination of rising costs, shortages of spare parts and poor maintenance and thus low vehicle utilisation, irregular services, over-

Table 5.2 Bus performance, productivity and cost data under different regulatory regimes in selected African cities

	Public operator			Private operator		
	Douala	Yaoundé	Dar es Salaam	Harare	Dar es Salaam[2]	Jos[2]
Performance and productivity						
Average fleet availability (%)	85	70	47	87	80	89
Trips operated to schedule (%)	89	81	86	77	N/A	N/A
Total passengers daily	248,000	170,000	220,800	350,000[1]	504,000	50,000
Passengers per bus operated daily	1,198	1,429	2,208	603	2,100	200
Passengers per km operated	7.6	8.1	10.1	2.6	10.3	1.2
Kilometres per bus operated daily	158	176	219	223	204	170
No. of breakdowns per 10,000 km	10.8	6.0	15	3.0	–	–
No. of accidents per 100,000 km	5.3	5.6	4	1.0	–	–
Staff per bus operated	5.8	10.8	13.5	3.7	3.0	1.0
Average passenger load[2] (km)	4.6	5.8	5.7	12.7	4.1	7.0
Average load factor[2]	0.3	0.4	0.6	0.6	0.8	0.7
Cost indicators (%)						
Staff/personnel	40.0	26.8	29.3	23.0	11.0	6.5
Fuel/lubricants	13.5	11.6	19.4	22.5	31.7	31.5
Tyres, tubes, spares and maintenance	15.0	14.8	16.3	40.8[3]	23.9	18.5
Depreciation/interest	20.2	24.1	9.0	5.8	28.8[4]	42.5[4]
Taxes/licences/insurance	4.3	3.0	1.2	0.7	4.6	1.0
Miscellaneous	7.0	19.7	24.8	7.2	–	–
Total of cost indicators	100	100	100	100	100	100

Notes: [1] Estimated from ticket sales   [3] Includes wages of maintenance staff
[2] Based on TRRL surveys   [4] Depreciation based on a capital recovery factor of 15 per cent for 5 years
Source: Maunder, D. A. C. (1990) *The Impact of Bus Regulatory Policy in Five African Cities*, Crowthorne: TRRL Research Report 294

staffing, high-level corruption, low staff morale, absenteeism and stagnant or falling revenues for various possible reasons. Reliance on subsidies has therefore increased. However, the ability of the public sector to maintain necessary – or even existing – levels of subsidy has declined almost everywhere as a result of the debt crisis and subsequent structural adjustment and economic recovery programmes generally instituted at the behest of the IMF, World Bank and bilateral donors.

One of the objectives of structural adjustment has been to reduce the direct role of the state in economic activity, restricting it as far as possible to facilitation and regulation of private sector activity. Accordingly, this has served as a means of promoting the Northern vogue of deregulation and privatisation across the Third World, often in the face of strong local opposition. Deregulation (also known as liberalisation) means the relaxation or removal of strict controls over the scope, quantity or quality of operations. By contrast, privatisation refers to the transfer of ownership and control from the public to private sector.

Municipal or parastatal bus operations have been a prime target of these policies, although the extent and speed of reform have varied widely. Full deregulation implies the removal of quotas, effective monopoly or oligopoly rights, the opening of bus services to any potential operators, be they public or private. Often, however, they have to comply with stipulated criteria regarding financial guarantees, minimum service levels and/or professional competence. Enforcement of such requirements is, of course, an entirely different matter. In reality, the experience of most cities has been mixed, with intermediate forms of deregulation and privatisation, and some benefits. This is elaborated in Case study H.

Case study H

The impact of regulatory change in Harare's bus sector, 1988–94

Figure H.1 shows the changes in regulatory policy and ownership in Harare, Zimbabwe, from 1988 to 1993. Clearly, ownership and the degree of regulation do not always change in the same direction. Research in Harare has shown that partial deregulation,

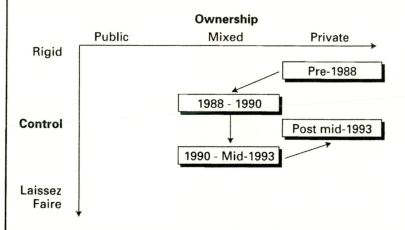

Figure H.1 Changing ownership and control of buses in Harare
Source: After Maunder and Mbara (1995), Figure 3

Case study H *(continued)*

coupled with direct government participation in the stage bus operations in which it became the majority shareholder in 1988, have led to improved services both of the stage buses themselves and of the bus service as a whole following introduction of 'commuter omnibuses', a type of minibus paratransit operating on a hail-and-ride basis. The total fleet expanded, so that the additional carrying capacity reduced passenger waiting times, while new routes were opened into areas not previously served. This in turn led to a 'redeployment' of emergency taxis (a form of paratransit shared taxi) to areas with inadequate public transport. 'Dead' or empty running by stage buses declined as new

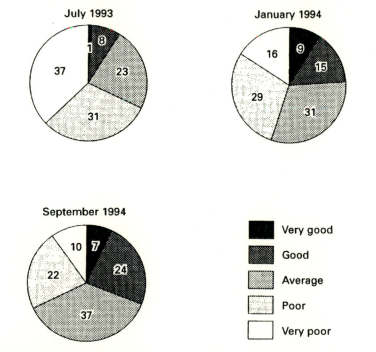

Figure H.2 Passenger opinions of changing bus service types and levels in Harare, 1993–4
Source: After Maunder and Mbara (1995), Figure 13

Case study H *(continued)*

Table H.1 Transport modal split, Harare, 1989–94 (%)

Year	ZUPCO stage bus	Emergency taxi	Commuter omnibus	Meter taxi	Motor car or cycle	Cycle	Walk	Other	Total
1988	18	7	–	0.5	30	1.5	42	1	100
1991	24	10	–	1	16	1	45	3	100
1992	31	9	–	1	17	5	36	1	100
1993	23	18	1	1	16	3	38	–	100
1994 (Jan.)	25	18	4	1	14	3	35	–	100
1994 (Sept.)	20	9	16	0.5	14	5.5	34	1	100

Note: ZUPCO = Zimbabwe United Passenger Company
Source: Maunder and Mbara (1995)

vehicles proved more reliable, while productivity (in terms of average kilometrage operated) per vehicle per day increased, as did off-peak bus availability; peak availability remained high. On the other hand, cost recovery of the investment in new buses would take time and require fare increases. Also, commuter omnibuses appear to be more accident prone than stage buses and to be contributing to congestion through their rapidly increasing numbers. As revealed in Figure H.2, passenger perceptions of the improvement were very marked, even over the short evaluation period following the introduction of commuter omnibuses.

When aggregate modal split in Harare is examined (Table H.1), it is clear that overall bus usage has increased markedly. Initially, the commuter omnibuses merely counteracted (or were offset by) the decline of stage buses, but by 1994, total bus ridership accounted for 36 per cent of the total. The role of emergency taxis fluctuated markedly from year to year, falling steeply from 1993–4 as commuter omnibus popularity rose. Equally significant, in the light of discussion earlier in this chapter, are the marked fall in the level of walking journeys and the rise in use of bicycles.

Sources: Maunder, D. A. C. and Mbara, T. C. (1993) *The Effect of Ownership on the Performance of Stage Bus Services in Harare, Zimbabwe*, Crowthorne: TRL Project Report 25; Maunder, D. A. C. and Mbara, T. C. (1995) *The Initial Effects of Introducing Commuter Omnibus Services in Harare, Zimbabwe*, Crowthorne: TRL Report 123; Maunder, D. A. C., Mbara, T. C. and Khezwana, M. (1994) 'The effect of institutional change on stage bus performance in Harare, Zimbabwe', *Transport Reviews* 14(2): 151–65.

Extreme privatisation has sometimes led to bus services being sold off lock, stock and barrel to private bidders – often large corporations or wealthy entrepreneurs with close political connections to the government concerned. At worst, this may result in little change other than the replacement of a public monopoly with a private monopoly. This and/or the withdrawal of previous subsidies may lead to increased fares and reduced service levels on quiet routes or at off-peak times, thus leaving passengers with inferior and perhaps less affordable public transport. A recent evaluation of total urban bus deregulation in Chile since 1979 suggested that the outcome was precisely the opposite of what had been predicted. The totally free access to the market and uncontrolled nature of fares has led to increased real fares and a less diverse service. This was attributed to the nature of urban transport supply rather than the actions of a cartel (Darbéra 1993).

Conversely, full privatisation and deregulation together have sometimes precipitated cut-throat competition, with excess capacity and unsustainably low fares resulting in considerable instability and even conflict between drivers of different companies. The point is that, while careful liberalisation and/or privatisation in accordance with local conditions may well lead to improved performance and a better overall public transport system, the widespread belief among donors that benefits are automatic and unqualified is simplistic, ideologically based dogma. As the Zimbabwean case study revealed, some forms of regulation and direct state involvement in transport provision can be perfectly compatible with improved efficiency and service levels.

Paratransit

As already observed, the various unconventional forms of vehicle-based service, known as paratransit or intermediate transit, comprise an important element of the urban public transport systems in many cities and towns. Table 5.3 provides details for a selection of large urban centres, suggesting that they account for anything from 20 to 60 per cent of motorised trips, with around 50 per cent being a typical figure. If non-motorised forms were included, the total paratransit percentage would be higher still.

Their popularity and importance stem from the inadequacy (in terms of areas covered and/or level of service) and cost of conventional public transport. Originally, motorised paratransits were often illegal, either because they involved conventional vehicles being used to provide

Table 5.3 Paratransit in selected cities of the South

City	Type	Fleet size	Share of motorised trips (%)
Delhi	Autorickshaws: converted scooters	28,000	20
	Pedal rickshaws	5,000	–
Istanbul	*Dolmus*: shared 5- and 7-seat taxis	16,000	} 50
	Minibuses	4,000	
Jos	Shared taxis	1,900	} 60
	Donfo minibuses	250	
Manila	*Jeepnies*: shared taxis (based on military Jeeps)	28,000	54
Nairobi	*Matatus*: mainly converted pickups	2,000	50
Surabaya	*Bemos*: 10-seater minibuses	3,200	30
	Becaks: tricycle pedicabs	38,000	–

Source: Armstrong-Wright (1993: 25)

unconventional services without licences, even if these were available, or because they were unconventional vehicles. The general attitude of governments and local authorities was negative, regarding them as a scourge on grounds of illegality, licence fee evasion, lack of quality control, poor maintenance and unsafe operations, high accident rates and their contribution to rising congestion. Operators were harassed, arrested, fined, and their vehicles sometimes impounded.

Gradually, however, and sometimes with the support of development agencies, attitudes have begun to shift. Their important contribution and the futility of attempts to suppress them are more often recognised. Sometimes, too, their profitability and the opportunities for capital accumulation thus offered make them attractive investments to political leaders, wealthy entrepreneurs and professionals. As a result, many local authorities or national governments have introduced measures to legalise or at least regularise their status to some extent. There has been some upgrading of, and investment in, new vehicles as a result. Minibuses, in particular, are replacing sedan cars, converted pickup trucks and other customised vehicles in many places (Plate 5.7). Such initiatives parallel the strategies behind aided self-help shelter construction or the encouragement of 'appropriate' forms of the so-called informal sector – of which paratransit is actually an example.

Consequently, paratransit operators may need to obtain licences and insurance cover, submit to some form of vehicle inspection and/or driver

Plate 5.7 Various types of *trotro* (shared minibus taxi) at the main terminus in central Accra

certification, and use specified ranks or termini. Driver/owner associations now exist in many cities and act as lobbies as well as support networks. Nevertheless, as with the Kenyan *matatus*, their status is often semi- rather than fully regularised. Safety records remain generally poor, overloading often severe and accident rates high and driving fast and reckless (Case study I).

Case study I

The *matatus* of Kenya

The history of *matatus* mirrors closely the general experience with motorised paratransits outlined in the main text. Their originally illegal status was ended by Presidential decree in 1973 but relevant legislation was passed only in 1984. Town planning mechanisms have still not really been adapted. The 1973 decree

Case study I *(continued)*

Plate I.1 Decoration of older *matatus*, such as this custom-built body on a light lorry chassis in Nairobi, is common

classified them as Public Service Vehicles (PSVs). However, an anomalous situation arose as they were still owned and operated as if they were private vehicles. There was still a lack of control, so licences, insurance and other safety requirements were lacking, accidents increased and police harassment was stepped up. The levying of fines provided a substantial source of income for the state, over and above bribes paid to police to avoid fines.

A comprehensive survey in Nairobi in 1982 found some 2,000 *matatus* in operation, divided equally between full- and part-time service. Over half belonged to fleets of at least two vehicles, but 35 per cent were owner-driven. Converted pickup trucks with a capacity of 18 (14 seated and 4 standing) were the most common design, although custom-built bodies and more recently also production line Japanese minibuses have become increasingly popular (Plate I.1). In towns and rural areas, pickups still dominate (Plate I.2). In Nairobi, *matatus* compete effectively with stage

Case study I *(continued)*

Plate I.2 Converted pickup truck *matatus* tout for passengers at the central terminus in Nakuru, Kenya

buses operated by the Kenya Bus Service (KBS), and account for a little over half of all motorised public transport trips. The percentage is likely to have increased somewhat since the early 1980s, on account of the losses, fleet shrinkage and other problems experienced by KBS.

Most *matatu* users in cities are low-income residents, as fares are lower and routes often more convenient than with stage buses. They are also more appropriate for people travelling with substantial luggage. However, as with many forms of paratransit, safety remains a major concern both to passengers and to local authorities.

Under the Traffic (Amendment) Act of 1984, *matatus* were fully legalised, requiring PSV licences, annual inspection by the police and passenger insurance for up to 25 passengers (the maximum permitted under the definition of a *matatu* in the Act). Drivers have to be a minimum of 24 years old and to have held a valid licence for at least four years.

Case study I *(continued)*

Plate I.3 *Matatus* often carry considerable freight and luggage in addition to passengers, frequently leading to overloading

Police enforcement, both in cities and rural areas is regular. Roadblocks are common on the outskirts of towns, and drivers of overloaded or otherwise irregular vehicles are subject to spot fines unless appropriate bribes are paid. This practice is widespread. Most rural *matatus* are overloaded to some extent (Plate I.3), especially on routes where other forms of public transport are lacking. Following the expulsion of non-Kalenjin ethnic groups from West Pokot district (along the Ugandan border) in the ethnic clashes of 1992–3 and 1993–4, the number of *matatus* in operation has declined, creating a local shortage, especially off the longhaul bus routes, where *matatus* are often the sole alternative to hitching or walking. The problem is most acute on market days, when people may walk tens of kilometres. In April 1995, I counted 32 people plus a considerable amount of luggage on one such old and battered converted pickup truck struggling over an undulating road. No fewer than 13 of the people were outside the vehicle,

Case study I *(continued)*

---

hanging in the open doorway at the rear, standing on bumpers and squatting amid the luggage on the roofrack.

Source: Lee-Smith, D. (1989) 'Urban management in Nairobi: a case study of the *matatu* mode of public transport', in R. E. Stren and R. R. White (eds) *African Cities in Crisis: managing rapid urban growth*, Boulder, Colo.: Westview.

By contrast, non-motorised forms of paratransit like rickshaws and cycle rickshaws are often traditional, with longstanding histories in particular cities, especially in South and Southeast Asia. They frequently provided – and sometimes still do – a substantial share of public transport, especially in central or older parts of large conurbations (Plate 5.8). They have much in common, in terms of characteristics of speed and trip length, with walking or bicycles as discussed earlier in this chapter, so tend to be slower than motorised alternatives and used principally for trips of up to about 3 km. Tricycles and trishaws occupy virtually as much road space as cars, leading to the accusations of congestion and inappropriateness that were discussed in the previous section of this chapter. As explained there, non-motorised paratransits consequently face pressure for their phasing out on practical 'town planning' as well as ideological grounds. However, there is a strong case for viewing them more positively as having a role within a diverse and multifaceted urban transport system, especially for low-income residents and tourists (Case study J).

Case study J

The *cyclos* of Phnom Penh

As a result of the destruction wrought by the murderous Khmer Rouge regime in Cambodia from 1975 to 1979, this Southeast Asian country is now one of the poorest in the world and is still struggling to rebuild its shattered economy and infrastructure. It is therefore not surprising that non-motorised public transport

---

Case study J *(continued)*

Plate J.1 *Cyclos* still form the backbone of Phnom Penh's public transport system, although the number of motorbike taxis (background) is increasing rapidly

remains far more important in Phnom Penh and other urban centres than in most other Third World cities. The principal form of such transport is the *cyclo* or pedal tricycle (Plate J.1).

In 1989, there were 9,662 registered *cyclos* in Phnom Penh, along with 4,078 motorbikes. Although there is strong demand for transport and perhaps another 1,000 unregistered *cyclos* exist, the Department of Municipal Public Transport has held the registered total of *cyclos* constant since 1989 in an effort to limit road congestion. By contrast, motorbike numbers have risen dramatically. Their use for commercial purposes has not been regulated up to now, so it is impossible to gauge the number being used as taxis. However, *cyclo* riders interviewed in 1992 claimed that the impact of increasing competition from motorbike taxis upon their own livelihoods is substantial, with earnings being reduced by up to 50 per cent.

This provides evidence of increasing competition from

Plate 5.8 Bicycles with sidecar and side platform being used to transport (above) a passenger and (below) fresh produce in Chinatown, Singapore

Case study J *(continued)*

motorised public transport – a universal experience for providers of non-motorised modes, the broader consequences of which were discussed in Chapter 3 – although it is unusual for this competition to come from motorbikes rather than buses or shared taxis. This reflects the still low, but rising, levels of at least some incomes in that, although motorbike taxis are quicker, have more prestige and cost more than *cyclos*, they are considerably cheaper than car-based taxis. Buses do not operate in Phnom Penh, despite its size (750,000 registered residents in 1989, with at least another 100,000 seasonal workers).

Cyclo riding provides important income-earning opportunities for some of the city's poorest residents. Eighty per cent of a survey sample were migrants from rural areas, compared with only 28 per cent of those engaged in other 'informal sector' work. They were all male, reflecting cultural constraints and divisions of labour which deem it inappropriate for women to undertake such work. Again, although Vietnamese, Chinese and other non-Cambodian ethnic groups comprise about 5 per cent of the country's population, and form a significant proportion of 'informal sector' workers, none was found in the survey of *cyclo* riders. This probably reflects their poor knowledge of the Khmer language and/or the city's layout.

The lack of capital greatly restricts the options of migrants: whereas 68 per cent of resident *cyclo* riders owned their vehicles, this was true of only 9 per cent of migrants. Instead they had to hire their vehicles, although this had the advantage of enabling a trial period for new entrants so that a permanent commitment did not have to be made immediately. In addition, hirers could store their *cyclos* safely while returning to their rural families at regular intervals. Family and other social networks based on area of origin or current urban residence are important for organisation of the *cyclo* industry and the provision of resources. Relatively few migrants saved up sufficient capital to buy a vehicle of their own.

Migration status proved the most important determinant of income-earning ability. Migrants generally worked longer hours for a lower income than residents; moreover, rental charges

Case study J *(continued)*

---

required a higher outlay than working capital and maintenance costs incurred by owner-riders. All registered riders had to pay for a rider's licence, *cyclo* registration, an annual road tax, a biennial roadworthiness test and a uniform, the combined cost of which could equal a week's income in 1992. Nevertheless, the incomes of *cyclo* riders were slightly higher than average for other 'informal sector' workers.

Source: Etherington, K. and Simon, D. (1996) 'Paratransit and employment in Phnom Penh: the dynamics and development potential of *cyclo* riding', *Journal of Transport Geography* 4(1): 37–53.

Mass rail transit (MRT)

The principal common feature of the many different types of railway system serving large urban areas is their ability to carry very large numbers of passengers, both per train and in terms of frequency of service. Three basic categories of rail transit system can be distinguished: conventional suburban trains, metros, and light rail transits (LRTs). While it is unusual to find MRTs and LRTs coexisting within individual cities, either system may be found in conjunction with conventional commuter railways. Each of the three types will be discussed in turn.

Conventional railways

Many cities rely in part or wholly on conventional railways linking the CBD (central business district) and industrial areas with some of its suburbs. At least some of the track used for suburban or commuter services is usually shared with freight and long-distance interurban services. Service frequency and the quality of coaches vary according to demand and local circumstances, although railways inherited from former colonial powers commonly still provide carriages in three classes. First class was reserved for European settlers, while the extremely rudimentary third class was for the poor. This hierarchy still exists in India, with class and caste having replaced race as the principal differentiators. In South Africa, third class, which was reserved exclusively for black and especially African commuters, has now been more or less phased out.

Very few new conventional urban railways have been constructed in recent decades, on account of the difficulties involved (not least the need to expropriate and clear broad swathes of valuable land in densely built up-areas), the unlikelihood of recouping heavy capital outlay through passenger fares, and the inflexibility of such systems. The principal exceptions are where large freight volumes are to be moved to and from new industrial and port complexes within metropolitan areas, and where new planned, high-density dormitory or satellite towns are constructed on the fringes of large cities. Even so, subsidies may be required, as in the branch line linking the vast new complex of Khayelitsha, built exclusively for Africans on the outskirts of Cape Town since the mid-1980s, with the existing suburban railway system.

As a result of the lack of new construction, coupled with very rapid rates of urban growth and sprawl over recent decades, existing conventional suburban railways may no longer serve major population concentrations within metropolitan areas, and/or be relatively marginal to the principal commuter corridors. Under such circumstances, which exist in Karachi and Lagos, for example, riderships are very low (from 7,000 to 15,000 passengers per day), and the operations totally uneconomical to maintain.

At the other end of the spectrum, however, well-located suburban lines may provide the backbone of commuter flows within a metropolis. Bombay provides an excellent example. The old parts of the city, CBD, harbour, inner suburbs, established high-income areas and the vast shantytown of Dharavi are all located on Bombay Island, which is connected to the newer parts of the metropolis on the adjacent mainland by only two main bridges. Each of these carries railway lines to the two central stations of Churchgate and Victoria, both constructed by the British in ornate Victorian styles. Up to 2 million passengers commute each way by rail daily, making this probably the busiest commuter railway system in the South. The eight- or nine-coach trains are in almost constant motion, with peak frequencies of 4–5 minutes on each route. They are overloaded to a remarkable degree for several hours in both mornings and afternoons during the working week (Plate 5.9). According to Armstrong-Wright (1993: 60), the trains are designed for crush loading of 1,728 passengers but regularly carry over 3,000. The

Plate 5.9 Passengers hang out of every door of a mid-morning Bombay commuter train into the city centre

average rail trip length in metropolitan Bombay is 24 km, four times the average bus trip.

Metros

Not least because of the limited opportunities to expand conventional suburban railways outlined above, it has become fashionable over the last twenty-five or so years for large metropolises in the South to invest in metros based on European designs and standards. Until very recently, the majority of these were in the NICs of Latin America, although the number in South and Southeast Asia and China is growing rapidly, a trend set to continue. However, metros are extremely expensive, with capital costs ranging from US$13–106 million per kilometre (Table 5.4), depending on terrain, subterranean rock conditions, labour rates, the density of the built-up areas served and the cost of property expropriation and station construction. As Table 5.4 also shows, most metro networks are predominantly underground,

Table 5.4 Characteristics of selected metro systems

City	Capital cost (US$m/km)	% under ground	Length (km)	Station spacing (km)	Minimum headway (min:sec)	Cars per train	Hourly design capacity (p/h/d)	Peak direction load (000/h)	Average journey speed (km/h)
Cairo	13	11	42.5	1.3	2:30	6/9	60,000	22	–
Calcutta	26	95	16.5	1.0	2:10	8	59,000	5	33
Hong Kong									
Kowloon	64	77	26.0	1.1	2:00	8	75,000 ⎫ 81		33
Island	106	84	12.5	1.0	3:30	7	38,000 ⎭		33
Mexico	30	75	131.0	1.2	1:55	9	46,000	65	35
Porto Alegre	10	0	26.7	1.9	6:00	4/8/12	16,000	11	41
Pusan	32	79	32.0	1.0	2:00	6	27,000	13	32
Rio de Janeiro	83	100	11.6	0.8	3:00	6	45,000	22	29
Santiago	36	81	26.0	0.7	2:40	5	20,000	20	32
São Paulo									
Line 1	80	82	17.0	0.9	1:45	6	58,000 ⎫ 57		29
Line 2	–	32	11.5	1.2	1:50	6	48,000 ⎭		38
Seoul	45	80	116.5	1.2	3:00	6	29,000	–	36.5
Singapore	37	30	67.0	1.6	2:00	6	–	14	–

Source: Fouracre *et al.* (1990); and Armstrong-Wright (1993: 41)

requiring costly tunnelling operations. Of the cities listed, only Porto Alegre, Cairo, Singapore and São Paulo Line 2 are predominantly or exclusively above ground.

In view of the cost, the export and installation of metro systems has therefore often been facilitated by guarantees, subsidies or loans from donor country governments and/or international development agencies. Indeed, such arrangements were sometimes decisive in winning the contracts against fierce international competition. In most cases, little if any attempt has been made to adapt the systems to local conditions. Given the scale, complexity and cost of these projects, international consulting firms and contractors have usually been responsible for successive stages of the project cycle, namely planning and design, construction, installation and management. Intensive training of local personnel has been necessary and some expatriate technical staff have commonly had to remain in post for considerable periods to continue with training and ensure adequate maintenance and safety. While these export orders have provided a substantial boost to the consultants and manufacturers concerned, the terms of these arrangements have prompted some observers and critics to question such apparent Western 'technological imperialism', especially in smaller cities where the justification for expensive, hi-tech solutions is particularly debatable (Dick and Rimmer 1986). The experience in Calcutta provides a notable antidote and exception (Case study K).

Case study K

Against the trend: the Calcutta metro experience

Perhaps the most conspicuous exception to the pattern of hi-tech, external dependence is provided by the Calcutta metro, designed and built almost exclusively by Indians with local technology and skills apart from the employment of international consultants for aspects of tunnel design. Whereas modern hi-tech tunnelling equipment was imported for use in most other metro systems, the necessary work in Calcutta was undertaken by cheaper, labour-intensive means, providing much-needed local employment, albeit at the cost of slower progress on a system which is 95 per cent underground. This approach also kept construction costs to about $26 million per km, lower than all the metros listed in Table 5.4, apart from Cairo and Porto Alegre (both predominantly above ground), and significantly below that for any other predominantly underground system.

However, major problems were encountered at various stages as a result of underground conditions, the risk of subsidence and tunnel collapse. Labour relations, funding and policy issues also proved problematic at times, contributing to substantial delays. Although construction began in 1973, it took eleven years before the opening of the first 10 km, comprising two separate sections. These will be linked by the final section of 6.4 km, due to open in 1995. At 65,000 per day, patronage in the first few years was dramatically below the design capacity of 59,000 per hour. This is partly because of the lack of networking and also on account of fares being double those on buses and trams. Revenue covered only 40 per cent of operating costs, let alone recouping the initial capital outlay.

Morever, a simplified (partial) economic evaluation undertaken for a comparative TRL study of metro performance suggested a negligible economic rate of return on the investment (between 1.0 and 2.8 per cent). This was the lowest of all the metros studied, and was attributed to a combination of 'low levels of patronage, low value of time and high capital costs, inflated in economic terms as a result of long construction delays' (Fouracre, Allport and Thomson 1990: 15). At the other extreme, high values of time,

Case study K *(continued)*

high patronage and/or quicker construction yielded remarkable economic rates of return of up to 20.5 per cent in Singapore, 18.5 per cent in Hong Kong and 16.8 per cent in Cairo.

In many ways, therefore, the Calcutta metro illustrates well the developmental issues and trade-offs involved in such systems. However, discussion of such points is conspicuously absent from the sources cited here. Conventional transport economists regard Calcutta's metro as a failure, with an inadequate rate of return to justify the capital outlay, and early revenues covering less than half the operating costs. Although some improvement is likely once the missing middle link section comes on stream, the overall picture is relatively unlikely to change dramatically. The project would therefore not have been approved by prevailing conventional economic yardsticks, and certainly not on the labour-intensive, relatively low-tech basis that construction was undertaken.

Economic viability might have been achieved using conventional international consulting and construction firms and hi-tech approaches; this would certainly have been the strong recommendation if the project were to be approved at all. However, as already indicated, this would greatly have increased capital costs, the drain on India's foreign exchange reserves (perhaps necessitating international loans), and the reliance on expatriates. At the same time, local employment generation, skills and technology utilisation, manufacturing capacity and the civic pride that self-reliance has produced, would have been dramatically reduced. Yet it seems that the path followed in this case, once the decision to proceed with a metro system had been taken, provided a more appropriate solution to the city's needs. It also serves as an example to others that, given a relevant degree of local technological and skilled labour capacity, external reliance and dependence are not unavoidable in the adoption of relatively sophisticated transport modes and solutions.

Sources: Fouracre, P. R., Allport, R. J. and Thomson, J. M. (1990) *The Performance and Impact of Rail Mass Transit in Developing Countries*, Crowthorne: TRRL Research Report 278; Armstrong-Wright, A. (1993) *Public Transport in Third World Cities*, London: HMSO.

Plate 5.10 Singapore's ultramodern Mass Rapid Transit (MRT)

As with conventional railways, the level of passenger use depends upon several related factors: the extent of the network, the extent to which it serves areas of high population density and major commuting corridors, the nature and degree of interchange and complementarity with bus or other feeder services, and the relative level of fares. Network length varies considerably (Table 5.4), with only a few having extensive networks. Moreover, most of the networks comprise only one or two lines crossing the metropolitan area.

Despite having the third longest network in Table 5.4 and being predominantly above ground, Singapore's MRT (Plate 5.10) falls into this category, with two lines. Moreover, its stations are rather far apart (1.6 km on average), thus not serving all residential concentrations along its route very conveniently. Although the MRT is well integrated with bus services at station interchanges, this is a major reason why ridership is comparatively modest, and far below what might have been expected. The principal design objective would therefore appear to have been to provide a fast and high-quality service linking the city centre with outlying dormitory areas rather than serving all neighbourhoods along its route. Fares are reasonable and the quality of stations and trains extremely high. It therefore seems plausible that national prestige and status in terms of joining the elite club of cities with a metro may have figured significantly in the decision to build the MRT and in determining the configuration of the network.

Peak hour design capacities in metro systems vary by some 450 per cent, from 16,000 in Porto Alegre to 75,000 in Kowloon, Hong Kong (Table 5.4). Even the lowest figure is therefore somewhat higher than the peak bus capacity along bus lanes and other enhanced routeways, as discussed in the previous section. However, Mexico City's metro carries the largest number of passengers per annum, with Seoul some 40 per cent behind and Hong Kong and São Paulo in third and fourth places respectively. These figures suggest that patronage is strongly and positively related to network length, with other factors being of secondary importance. Another relationship revealed in Table 5.4 is between average journey speed and station spacing: those systems which operate at the lowest average speeds, notably Rio de Janeiro and São Paulo Line 1, have the shortest distances between stations. Conversely, Porto Alegre has the highest average speed and the longest station spacing. Most metro systems provide average journey speeds of 30–35 km/h.

In an attempt to maximise ridership and cover operating costs – and repay some of the capital costs of construction – many transit authorities initially sought to reduce or eliminate competition from bus services. With very few exceptions, most notably Mexico City and Tunis, however, such measures have failed as a result of opposition from bus operators and the public. Such policies are now generally frowned upon; the emphasis is on deregulation and competition enhancement. On the other hand, fare integration – the issuing of a single ticket valid for a journey involving bus and MRT in an effort to encourage use and avoid intermodal competition – has seldom been implemented, let alone proved effective. The principal exception is São Paulo, where this has been successfully achieved but only by means of a continuing operating subsidy to the metro (Armstrong-Wright 1993: 41).

Finally, it is worth mentioning that, as elsewhere, metros in the South are invariably owned by the national, regional or local state, although separate parastatal operating companies have sometimes been established. These may be restricted to the metro alone, or have responsibility for other modes of mass rapid transit (commuter railways, LRTs, buses) as well. While often subject to state policy constraints and, very importantly in view of widespread operating losses, also subsidies, the idea is to promote independence of management. Under structural adjustment, economic recovery and other recent policies, the trend has been to liberalise operating environments and policy constraints, with permissive or positive attitudes towards increased competition from buses and paratransit. Of course, metros and LRTs are, almost by definition, in a monopoly situation for their specific modes within individual cities.

Light rail transits

This category of mass transit comprises a spectrum of conventional tram systems operating in mixed traffic on rails embedded in (and flush with) the roads and more modern, innovative systems designed for much higher passenger flows and avoiding direct conflict with road vehicles. These tend to be hybrids, sharing some features of metros, some even being known as LRT metros. LRTs are electrically powered, obtaining their input from overhead cables. Whereas longstanding trams (such as in the Victoria district of Hong Kong) may have a maximum capacity of 6,000 passengers in each direction, the ceiling for some new systems is as high as 28,000. As indicated above, some LRTs exist in cities which also have a metro (like Cairo and Calcutta) but it is more common for one or the other to exist. Interestingly, the tram networks in these two cities are the two most extensive in the Third World.

In contrast to metros, LRT systems operate at low speed, using single or double cars, which are of either 4-axle rigid or 6-axle articulated design, and with a capacity of up to 200 per car. Whereas metros are co-ordinated and controlled from remote command centres, even though each train has a driver, LRTs rely on exclusive driver control. This is necessary on account of their operation in often heavy mixed traffic on road surfaces. Naturally, this is one of the principal reasons for their low speeds, which average from 8 to 12 km/h, and the unpopularity of traditional tram systems. LRT metro systems attain higher speeds (up to 25 km/h) and passenger flows (up to about 30,000/h/track, with up to 900 per train), on account of their segregated tracks and higher capacities. Inevitably, however, such innovations increase costs. Armstrong-Wright (1993: 51) cites figures ranging from US$300,000 per conventional tram car purchased from Central/Eastern Europe to US$1.5 million for hi-tech Western LRT cars, with new tracks and electricity transmission cabling costing around $4 million/km. For obvious reasons, segregated tracks also add substantially to costs.

Category	Description	Diagram
A	Fully segregated, not driveable by traffic	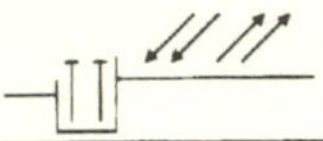
B	Segregated by kerbs, cars can use in an emergency	
C	Segregated only by white line and police control	
D	Not segregated, traffic has one lane	
E	Fully mixed, trams and traffic share same space	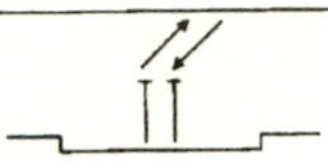
F	Pedestrian street, trams and pedestrians share same space	

Figure 5.3 Road and track design options for tram and LRT systems
Source: After Gardner *et al.* (1994), Figure 4

Figure 5.3 summarises the main options in road and track design for LRT systems, which range from full segregation (category A) to fully mixed vehicular traffic (category E) and fully mixed traffic and pedestrian streets (category F). The principal trade-offs are between cost of construction and maintenance for more segregated systems versus operating speed, safety and convenience. Whatever the design, there is also a trade-off between the time set between trams/trains (known as 'headway') and safety. Beyond a certain point, however, lengthened headways also lead to passenger crush at stops or stations, with safety risks and delays.

A comprehensive recent evaluation of LRTs found that the older tram systems in cities like Cairo, Alexandria, Calcutta, Dalian and Hong Kong, some of which actually predate the motor car, are slow, providing low passenger flows (no more than 6,000/h/direction), poorly maintained, and suffering network cuts (Table 5.5). At the other extreme, the

Table 5.5 Estimated performance of tram and LRT systems

System	Examples	Observed capacity (p/h/d) and speed (km/h)	Estimated capacity (p/h/d)
Tram – high proportion of street running	Alex-Madina Calcutta	6,000/6 3,500/12	8,500
Tram – high proportion of segregation	Cairo (Heliopolis) Tunis	12,800/14 9,300/16	13,000
LRT – grade separated 2-car trains 3-car trains	Manila –	19,000/30 –	19,400 26,700
Metro high output low output	Hong Kong Santiago	80,000/33 20,000/32	
Busway transit high output low output	São Paulo Ankara	19,000/20 7,300/13	14,900–27,900 5,800–18,100

Source: Gardner *et al.* (1994)

most successful systems in terms of passenger speeds and flows are dedicated, segregated modern (and far more costly) systems, of which the Manila LRT is the best example. It operates on raised concrete viaducts, and has many features of full metro systems. Peak hourly passenger flows of 19,000 per direction were recorded using two-car trains; this could be increased to 26,700 with three-car trains (Gardner, Rutter and Kukn 1994). The study concluded that, unless such systems could be afforded and were locally appropriate, then bus lanes or dedicated busways were a far superior option to conventional trams, unsegregated or semi-segregated LRT systems. They provide greater flexibility, higher flows, less overall delay, cost far less and can be introduced with relatively little disruption.

Conclusions

This chapter has surveyed the plethora of modes and forms of urban transport, highlighting their respective features, their interrelationships and the potential for contributing to a more effective and appropriate urban transport system. It is clear that no single form of urban transport can exist in isolation, and that present, often fragmented, arrangements and policies can be improved.

Among the principal challenges facing transport planners and urban managers in large cities of the South are the *rate* of population growth and scale of urban expansion. A high proportion of new rural–urban migrants are poor, with the result that they have very limited means with which to meet their urban transport needs. For low-income residents, in particular, walking remains probably the most important form of movement, although average walking trips are relatively short. The reliance on walking and cycling declines with increasing income levels.

The available evidence also shows clearly that, where the urban incomes of at least a significant proportion of urban dwellers are increasing, as in Latin America and much of Southeast Asia, ownership of private means of transport, especially motorbikes and cars, has increased, again often at very rapid rates, with resultant increases in congestion and pollution. A recent study of environmental problems in metropolitan Bangkok highlights traffic and transport (and their energy consumption) as one of the major areas of concern:

traffic conditions have deteriorated rapidly in recent years, and are now arguably the worst of any large urban area in the world. In 1989 the average travel speed on main roads in the Bangkok metropolis was 8.1 kilometres per hour . . . during peak hours. . . . Traffic has slowed since then, and is often slower than the projected 2006 travel speed of 4.8 kph. . . . The principal causes of this deterioration in conditions are the rapid growth and unrestrained use of private vehicles; a poorly developed road network; lack of investment in the current mass transit system (i.e., the public bus fleet); and poor planning and government indecision.

During the 1978–1991 period, the number of motor vehicles registered in the BMA [Bangkok Metropolitan Authority] alone increased from 505,000 to over 2.6 million, an annual average rate of 13.5 per cent. . . . While an average of 446 vehicles were added to the BMA motor vehicle fleet *every day* during this period, future projections are even more ominous, as the phase of large absolute growth of the motor vehicle fleet is only just beginning. By 2006 the number of private vehicles may increase by three to four times the 1991 level. By comparison, while the number of motor vehicles increased by 250 per cent during the 1978–1988 period, the number of buses owned or franchised by the Bangkok Mass Transit Authority – the only form of organised mass transit currently available in the Bangkok metro area – increased by just 6.7 per cent. . . . Moreover, despite the low level of bus fleet growth during this period, the average number of passenger trips per day increased by 84.7 per cent.

(Setchell 1995: 7–8)

The pollution and waste of energy represented by vehicles idling in traffic jams in metropolitan Bangkok has a direct cost of millions of US dollars, quite apart from the indirect costs and the time and productivity lost through the congestion.

While perhaps more extreme than in other large metropolises, the difference is merely one of modest degree. Peak travel times ('rush hours') are known to be extending through both the morning and after-noon in many such cities, and chronic traffic congestion is commonly now the norm rather than the exception during weekdays. Environmen-tal costs, not least to human health, are still generally regarded as of secondary importance (at best) relative to economic growth and increas-ing incomes by Third World governments, but in some NICs suffering from acute pollution and/or traffic congestion, increasingly strict poli-cies to address the problems are being introduced and enforced. For example, Singapore and Hong Kong were the first in the world to introduce area licensing schemes to restrict car numbers in their CBDs. High vehicle duty and licence fees have also been introduced, while experiments with electronic road pricing (sophisticated and auto-mated forms of tolls) have been under way for some years.

Moreover, these problems highlight the importance of appropriate and affordable public transport systems. The diversity of public transport, embracing paratransit with taxi-like and bus-like characteristics, various bus types, and rail-based systems has been surveyed, providing insights into their respective features and the wider developmental issues asso-ciated with different state attitudes and policies to them. In general, governments still overwhelmingly tend to favour hi-tech, modern approaches, regarding traditional forms as outmoded and even problem-atic. The modernising elites retain considerable effective power. As Dick and Rimmer (1986: 195–6) observe,

Foreign consultants seem to have been instrumental in the adoption of advanced capitalist technology, initially in the form of stage buses but increasingly in the form of ultra-modern rapid transit systems. The obsession of national bureaucrats that their cities should have the image and trappings of a modern city in an advanced capitalist country, involves a strong element of imitation. Moreover, Third World governments have wide discretion in the countries from which they can enlist aid for urban public transport development, in the consultants whom they employ, and in the recommendations which they finally accept. This discretion is enhanced by competition between advanced capitalist governments, consultants and suppliers of technology for valuable contracts in the Third World. Mere observation of 'technological dependency' therefore does not establish cause and effect. Thus, in the field of urban public transport at least, neither dependency theory nor modernization theory seems altogether relevant. As is so often the case, the truth seems to lie somewhere in between.

Key ideas

1 Urban transport systems are extremely diverse, including both motorised and non-motorised modes, although these are seldom adequately integrated and the problems of congestion and pollution are increasing in large metropolises.
2 Although non-motorised forms of paratransit often have long traditions and may serve the needs of many poor residents while providing many jobs, governments are generally still seeking to phase them out on grounds of congestion, slow speed and claimed inappropriateness in modern cities.
3 Private vehicle ownership rises rapidly with increasing average incomes or greater prosperity for significant segments of urban populations. Motorbikes are cheaper, use less road space and may generate less pollution, but are regarded as inferior to cars, which are the principal contributors to rising congestion.
4 Public transport often suffers from inadequate investment and maintenance, and restrictive policies which limit the scope for innovation and the appeal of buses or trains to the public. Recent policy initiatives have sought to liberalise operations, enhance efficiency and reliability and maximise the complementarity between different modes of transport, both public and private.
5 Many city governments have invested in relatively hi-tech and expensive mass transit systems, both metros and LRTs, in an effort to enhance the modern image of their cities. Sometimes these are appropriate, but there is much evidence that hi-tech is not necessarily best (as demonstrated with respect to the Calcutta metro), and that, for example, carefully designed bus lanes offer higher passenger capacities than most LRT designs for a fraction of the cost and without buying new technologies.

6
Technological change and intermodal competition in long-distance transport

Introduction

In Chapter 2, some of the broad issues relating to the impact of technological change in international transport were introduced, with reference to the examples of containerisation in the sea freight market and safety standards in international civil aviation. Not only do such changes have potentially significant employment implications (usually negative because of the labour-saving nature of much new technology) but the investment costs may be substantial. Moreover, poorer countries frequently have little choice in the matter because Northern countries adopt these technologies and international organisations dominated by them set safety or technology standards – with the best of motives – which preclude gradual or partial adoption. If poor Southern countries wish to participate in the respective freight and passenger networks at all, they need to follow suit to a large extent. In the case of unitised sea freight, for example, the only real element of choice would be whether to containerise all ports or only the principal ones, which would then have to act as transshipment and consolidation points for goods from other sources. Such a strategy would add to freight costs and delivery times, perhaps disadvantaging goods thus shipped relative to competing imports and exports from other sources.

This chapter explores the development issues arising out of such situations, the associated policies and the intensifying competition between transport modes for long-distance domestic and international transport of both freight and passengers.

Civil aviation

Although this mode of transport remains irrelevant to most poor people in the Third World, the extent of civil aviation has been growing rapidly in most parts of the world, notwithstanding some cyclical fluctuations (Chapter 2). Nevertheless, domestic air traffic generally comprises a negligible proportion of the total recorded at Third World gateway airports. The principal exceptions are a small number of NICs and potential NICs (e.g. Brazil, Mexico, South Africa, Iran, India, Pakistan, Indonesia and China) and countries like Saudi Arabia and Nigeria, where oil revenues, physical size and/or inhospitable terrain, and the existence of several large cities have contributed to rapid increases in domestic civil aviation since at least the 1970s.

The importance of international passenger and freight flows reflects the extent of tourist and business travel and the trade in high-value (often perishable or fragile) cargo. In order to maintain and enhance these links, the airports and associated infrastructure, and the quality of service on the national airline, must nowadays be as attractive and trouble-free to users as possible. Together with the safety standards imposed by international treaties, this implies that considerable investment in facilities and skills development is necessary.

To return to the example in Chapter 2, the degree of choice may be very limited, e.g. reducing the capital outlay and the level of skilled personnel required by equipping airports for daylight or manual aircraft movements only, rather than installing sophisticated automated radar and so forth. However, this might reduce the numbers of tourists to the country or the number of airlines using the airport as a stopover, if it creates inconvenience for airline schedules, which are usually based on preferred departure and arrival times in the North. International competition today is so intense that few countries are willing to take this risk.

So, for example, Kotoka International Airport in the Ghanaian capital, Accra, which had long been outdated, in need of improvement and expansion to cater adequately for current traffic volumes, finally underwent a major upgrade costing US$53 million on completion in 1993. The existing terminal had been built in the 1960s, shortly after independence. The country's prolonged political and economic upheavals had precluded any renovation happening earlier; traffic volumes were in any case more modest during the long recession and the initial years of structural adjustment. The overhaul, making Kotoka one of the most modern in the region, was designed explicitly as part of a strategy to promote international business and tourism. However, the Ghanaian government could not afford so large a capital outlay. In a deal typical of such arrangements, the government contributed the local labour costs, while $15.3 million was provided by the UK Overseas Development Administration and the balance was put up by the UK government's Export Credit Guarantee Department. The exports thus covered will need to be paid for by Ghana in due course; the opportunity cost – what could otherwise have been done with these resources – is clearly significant.

Third World airlines often have difficulty in competing with major international carriers based in North America and Europe which have vastly superior resources and market shares. Undercutting the price is one of the few options available, especially as this advantage to passengers frequently has to be traded off against less ostentatious service, older aircraft and more stops *en route*. The cost of new intercontinental jet aircraft is excessive for many smaller Third World airlines. They therefore either incur heavy debts through large loan agreements or resort to purchasing second-hand airliners and increasingly also to leasing agreements. During the long recent recession, resale prices fell and more attractive leases were available.

Although many national airlines now operate at least some wide-bodied jetliners, their average age is greater than in the fleets of leading world airlines. When wide-bodied aircraft first became popular, the ageing Boeing 707s and DC8s, for example, sold off by major carriers, were often bought by Third World airlines and air freight charter companies. The same is true of Boeing 727s and DC9s, now largely replaced by B737s, MD80 and Airbus series planes in the North. Even the occasional geriatric Caravelle can still be seen at airports in Africa and South Asia. These aircraft may need increasingly expensive maintenance as they near the end of their airworthy lives; they are also less fuel-efficient and more noisy than more recent generations of aircraft. This in turn has created problems for the airlines still operating them, as the majority of Northern airports have imposed noise restrictions in recent years. Hence, Boeing 707s and DC8s could no longer fly into major European and North American gateways unless modified with costly special 'hush kits' to reduce their noise and exhaust emissions. These are further examples of technological and regulatory change having unequal impacts and affecting poor countries' airlines severely. While a few have complied, and were still flying B707s into London's Heathrow Airport in mid-1995, for example, many other operators ultimately began phasing these aircraft out of service and updating their own fleets rather than incur major new expenditures on ageing jetliners. Today, the airlines of most countries in the South contain a range of short- and longhaul aircraft, including at least some wide-bodied planes with capacities of 200 passengers or more.

The proliferation of wide-bodied jets has certainly increased substantially the numbers of tourists and other international visitors travelling to countries in the South by air. Although the aggregate benefit to host countries of tourism is debatable, especially in the case of poor states where the import content of the industry is often high, most governments and the relevant private sector interests have consciously promoted tourism as a form of economic diversification and employment creation. In the age of the jumbo jet, numerous dedicated tourist zones have been developed, catering principally to mass intercontinental tourism with attractive images of sun, sea, sand and sometimes sex. Examples include the Bahamas and Barbados, Jamaica's Montego Bay, Goa in India, Phuket Island in Thailand, the beach hotel strips in Mombasa and Malindi in Kenya, and the equivalent outside Banjul in the Gambia and Freetown in Sierra Leone. In many Caribbean and other island states as well as popular beach and safari destinations like Kenya and Tanzania, tourism has become the principal foreign exchange earner.

Another strand of economic diversification facilitated by wide-bodied jets and the associated improvements in rapid, long-distance refrigerated transport is the international trade in fresh vegetables, soft fruit and cut flowers. These commodities, such as green beans, snow peas, mangetout, strawberries, mangoes, paw-paws, avocados and other 'exotic' tropical or subtropical fruit, and cut roses, lilies, orchids and other flowers, are now grown, transported to the airport, air freighted to Europe and North America and auctioned or sold on to distributors in large quantities. These are highly specialised and controlled operations, requiring skills and strict quality control to comply with quality and freshness requirements in the importing countries. The time taken from picking or cutting to purchase in the supermarket or florist on the other side of the world is often no more than 24–48 hours. Such operations are commonly undertaken or controlled by transnational corporations, thus ensuring quality and efficiency from the perspective of the North but arguably reducing the overall benefit to the growing countries. Import content may be significant, as is often the use of expatriate staff in senior positions. On the positive side, export revenues are boosted substantially by these new commodities, not least because they are comparatively high-value products and because the value-added in the producing country is greater than for conventional agricultural produce. Furthermore, the diversification of export crops reduces the exporting countries' vulnerability to fluctuations in export earnings.

Very few countries have felt able to neglect their civil aviation industries, as the employment potential of civil aviation is considerable. This includes direct employment in construction of the airports, maintenance of aircraft and ground facilities at airports, and airline personnel in the air and on the ground, as well as indirect employment in supplier industries (e.g. airline catering) and those serving travellers in hotels, restaurants and other tourist facilities. For similar reasons, the environmental costs, both direct and indirect, are seldom taken into account. While air movements remain at modest levels, this is understandable, but the impact of air and noise pollution around major hub airports is now considerable, not to mention the vast volumes of waste generated by the disposable crockery, cutlery, drinks containers and wrappings used by passengers.

Water and maritime transport

River and inshore zones

Traditionally, many inland communities around the world have relied on water transport, often using simple but appropriate technology, such as canoes and rafts made from reeds, hollowed-out logs and animal skins. Some are still in use today, principally on inland waterways. However, modern rowing boats and motor boats of various designs have become widespread, even in relatively remote areas.

Historically, maritime transport developed in many different regions of the world and, indeed, until the fourteenth or fifteenth century, Western Europe lagged behind most ancient civilisations in terms of such technologies. Thor Heyerdahl has ably demonstrated how long-distance trade across the Pacific, Indian and Atlantic Oceans was possible in remarkably sophisticated boats constructed from reeds and wood. The sailing prowess of the ancient Egyptians and Romans, not to mention the Chinese or Polynesian and Melanesian societies, is well established. Chinese and Arab seaborne traders had been active along the East African coast long before the arrival of the first Europeans, for

example. It was really only with the voyages of exploration and con-
quest which paved the way for European world domination that European shipbuilding technology gained the ascendancy.

There is a wide range of water transport forms in use today. Rowing
and motor boats of various designs provide the principal livelihoods of
coastal communities engaged in fishing and trade (Plate 6.1). In some
cases, villages have specialised in one such activity, providing the basis
for successful collaborative local industries (often with an export
component) and good standards of living. However, progressive
encroachment of foreign ocean-going 'industrial' fishing boats with
sophisticated sonar and other equipment and vast nets into the traditional inshore waters on which such communities depend, has undermined their livelihoods and threatened the sustainability of fish
resources. Two well-publicised recent examples are Kerala in India
and Senegal in West Africa.

Many inland and coastal communities also utilise small river boats to
send agricultural produce to ports for sale in local markets or transshipment into larger, ocean-going vessels (Plates 6.2 and 6.3). Their ability
to do so depends on the navigability of rivers and the appropriate

Plate 6.1 Inshore and sea-going fishing boats used in the Malaccan Straits off
the coast of Johor State, Malaysia

Plate 6.2 A tug and lighters at a groundnut collecting station on the Gambia River

Plate 6.3 Traditional canoes which bring palm oil from small inland collecting stations to the port of Calabar, Nigeria

location of a port. Transshipment of both imports and exports between ship and shore by means of small boats with shallow draught, known as lighters, is also important where large vessels cannot reach the quayside or jetty on account of shallow water or dangerous reefs. While such activities are slow in comparison to bulk or containerised loading, as only small quantities can be moved at a time, they do not require sophisticated technology and provide significant employment (Plate 6.4).

However, intensifying international and intermodal competition in terms of speed of delivery and service quality, coupled with the difficulties of transshipment out of or into containers and bulk cargo vessels, has tended to concentrate seaborne traffic increasingly in larger ports with at least some container or bulk loading facilities, utilising road or rail transport instead of lighters. Ports lacking such facilities, and until recently reliant upon palletised loading by crane, have had to modernise or face a substantial decline in traffic (Plate 6.5). The unloading of unusual or heavy cargoes required for development projects can still cause difficulties in smaller, poorly equipped ports (Plate 6.6).

Plate 6.4 Traditional lighters in the port of Malacca, Malaysia

Plate 6.5 Loading cocoa by conventional labour-intensive methods and mechanised conveyors, in Tema, Ghana

Plate 6.6 Unloading earthmoving equipment and other heavy cargo for development projects presents a substantial challenge in ports with limited facilities

Maritime shipping

Ocean-going vessels provide the backbone of international and intercontinental trade for all countries, even those which are landlocked, lacking their own coasts and seaports. These are dependent on transshipment and transit traffic through their coastal neighbours or up navigable rivers. This situation tends to increase their transport costs and render them vulnerable to pilfering, delays and diversions, as their freight may be regarded as of less importance than consignments for the countries in which the port and transshipment points are situated. This is particularly likely where port or other infrastructural capacity is limited. Zambian freight has often been subject to precisely these pressures over the last twenty or so years in the Tanzanian port of Dar es Salaam, which has suffered from poor management, obsolescent equipment and further limitations imposed by the shallow channels in the river mouth, which preclude large ships from docking. In times of political hostility, undue delays and even interference with transit traffic may be experienced because of border closures or freight diversion.

The importance of maritime transport has not diminished with changing technology, although this has altered its precise nature and the flows involved, as well as the relative importance of different origins and destinations. Indeed, the progressive globalisation of production and consumption, often controlled by transnational corporations, has increased the reliance of almost all countries on imports and exports, as overall self-sufficiency has generally declined in favour of greater specialisation in trade. The changing international divisions of labour have promoted greater diversity in the character and nature of specific trade flows. Deindustrialisation in former advanced industrial states in the North has meant that they now import many of the products and commodities that they previously exported, although they may export specialised high-value components and new products. Conversely, the most successful Asian NICs, like Singapore, Hong Kong, Taiwan and South Korea, are resource-poor and have built their economies upon the import of raw materials and the export of increasingly sophisticated consumer durables and capital goods around the world. At the other extreme, many poor countries, especially in Africa and the Caribbean, remain principally exporters of primary commodities in raw or semiprocessed form and importers of the bulk of their industrial goods.

The most significant changes in maritime technology since the 1960s have been containerisation of general cargo (often called unitised freight) and the development of very large bulk carriers for bulky, relatively low unit-value commodities such as metal ores, coal, crude oil and cereals. Each of these has required the construction of dedicated berths or terminals, with specialised loading and unloading equipment. In the case of containers, the key requirements are rolling cranes of different design from, and far higher lifting capacity than, conventional dockside cranes, and a large flat area for the stacking and rapid sorting of containers. This has been difficult to provide along existing quays and berths because of the need to reinforce the quays and to remove redundant warehouses. Certainly, the numerous and narrow finger quays which existed in many ports are totally unsuitable.

In the vast majority of cases, therefore, new berths or even dedicated new docks within harbours have been constructed at considerable capital cost. In some smaller ports, especially where the draught of ships able to dock is limited, containerisation has not been possible, or only on a very modest scale with the use of small coasters which transport the containers to larger ports for transshipment to intercontinental services. This usually has the effect of adding to transit times and costs, thus further disadvantaging and even marginalising the small 'feeder' ports and services from them. Conversely, container traffic tends to concentrate in a limited number of high-volume ports which serve a wider catchment than previously. Singapore provides the best example of this phenomenon, with the combination of transshipment and local import–export trade making it the world's busiest container terminal in terms of TEUs (twenty-foot equivalent units) handled (Case study L). Many high-income countries in the North have established inland container terminals as collecting points for densely populated and industrial regions. The containers are then transferred to coastal ports by fast, dedicated train services.

Case study L

Integrated transport, regional development and the Singapore Extended Metropolitan Region

Singapore has emerged as a leading entrepôt, manufacturing and financial centre serving regional and increasingly also global markets. Its strategic location on the Malaccan Straits, at the

Case study L *(continued)*

centre of a rapidly developing axis of the Asia–Pacific Rim and on the air route between Australasia and Europe, has facilitated the rapid development of this city state of 3 million people. Conscious and far-sighted government policy sought to produce a highly educated and skilled workforce able to compete successfully against higher waged labour in traditional industrialised states of the North. Against a background of authoritarian rule and direct state involvement in the economy, Singapore progressed from a short-lived import substituting industrialisation policy to one based primarily on exports.

It became a leading NIC, one of the four so-called Asian Tigers, and one of the leading world centres for the assembly of electrical and electronic goods and other manufactures. Diversification has seen the emergence of a major tertiary sector, specialising especially in finance and business services for Southeast Asia and beyond. Along with Hong Kong, it is the leading financial centre in Asia, increasingly forming part of the small group of 'control centres' for the world economy. Agriculture has shrunk in extent and scope, again as a result of deliberate policy, specialising increasingly in aquaculture, biotechnology and similar high-value forms of production.

The success of such policies has depended on two key elements: first, the ability to 'export' lower value agricultural production (e.g. the keeping of livestock and crop growing) and more basic manufacturing plant to neighbouring territories where labour costs are lower and pressure on land less severe but which could still be accessed and controlled readily from Singapore. The principal sites of such offshore production have been the city of Johor Bahru – the capital of Malaysia's southernmost state, and its hinterland, which are just across the short causeway on Singapore's northern coast – and Indonesia's Riau Island group just to the south. The largest and most important of these are Batam and Bintan (Figure L.1). These adjacent territories have become increasingly incorporated into what amounts to an extended metropolis, of which there are several emerging in Asia. These metropolises (*desakota*) comprise a functional economic area

Case study L *(continued)*

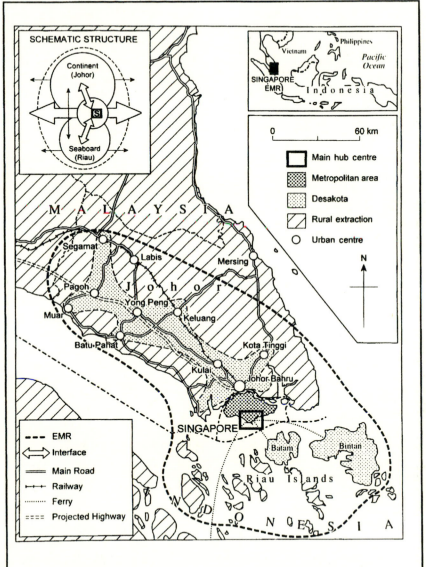

Figure L.1 The Singapore Extended Metropolitan Region
Source: After Rodrigue (1994)

Case study L *(continued)*

stretching far beyond the built-up metropolis at their core, including
smaller cities and towns and even villages, in which increasingly
specialised industrial production takes place. Like the associated
agricultural and recreational zones, they are intimately bound up
with the metropolis in each case. They extend the geographical area
over which population and production are spread, reducing to some
extent the inward migration or commuting pressure on the core area.

On the other hand, the necessity for good and reliable freight
and passenger transport is increased, both internally and externally. This comprises the second key element necessary for
success. Singapore has again proved extremely successful on
this score. The proximity of Johor and the Riau Islands has
enabled the establishment of fast, frequent and reliable transport
links within the extending metropolis, thus ensuring the integration of its various components and sustaining its expansive
momentum. Johor and Singapore are linked principally by road-
based transport, although small-boat traffic is also evident. Plans
for a new, higher capacity causeway have been drawn up. Across
the Malaccan Straits, fast and regular shuttle ferry services have
been introduced, linking Batam with Singapore in 30 minutes.

External transport has provided the springboard for Singapore's
phenomenal growth and development, which continued apace
through the 1980s, despite the effects of recession. The harbour
handles more ships now than even the famous Europoort in Rot-
terdam, and its container terminal has the highest TEU throughput
in the world (76,631 in 1990, compared with only 12,550 in 1980).
In 1990 a total of 106.2 million tonnes of freight were unloaded and
81.57 million tonnes loaded. The equivalent tonnages for 1980
were 50.12 million and 36.18 million respectively. In other
words, the number of TEUs increased almost sixfold over the
decade, while the tonnages of freight unloaded and loaded both
more than doubled. Japan and Malaysia are Singapore's largest
trading partners, while nearly 59 per cent of total 1990 trade was
with Asia as a whole. Given the nature of the extended metropolitan
region, trade with Indonesia is also rising substantially.

As discussed above, Singapore is now a significant owner of

Case study L *(continued)*

maritime vessels, mostly registered in Panama, Liberia and the Bahamas. In air transport, Changi Airport is the hub for many Asian and Australasian services as well as the home of Singapore Airlines, which boasts one of the most modern fleets in the world, among the largest in Asia. Until their recent shift to Bangkok, Qantas Airlines of Australia used Changi as its Asian hub for its services to different parts of Australasia from Europe. Although Hong Kong has the busiest airport in the South, in terms of passenger and freight throughput, Changi is not far behind, with over 48,800 aircraft landings in 1990. These brought over 7.2 million passengers in and 7.17 million out, while another 1.22 million passed through in transit. Over 324,000 tonnes of air freight were unloaded and almost 300,000 tonnes loaded. Despite strong competition between Singapore and Hong Kong, air and sea traffic between them is substantial, not least because Hong Kong lies on the principal sea route between Singapore and Japan. Rodrigue (1994: 72) concludes that

Transportation is a key factor related to the four processes of territorial development in the Singapore extended metropolitan region – densification, dissemination, extension and contraction. Through densification, a process of spatial accumulation of economic activities which aims to increase productivity, the territorial function of Singapore changes towards new sectors of activity. Dissemination, a spatial relocation of economic activities towards productive areas, is the process underlying the industrial development in Singapore's peripheral areas. Extension, a space/time collapse within transportation systems, enables economic activities to develop within the EMR while maintaining economies of scale and low distribution costs. Contraction, a rationalisation of distribution systems, is imposed by Singapore transportation firms facing growing transportation costs and competition.

A multimodal transport system, with efficient intermodal interchanges, is one of the main challenges facing the area in the near future, as each mode still functions predominantly separately.

Bulk carriers are very large, with deep draughts. They therefore require deep-water berths and also incur high port charges. In order to facilitate the quick turnaround on which their profitability thus depends, loading and unloading must be rapid. This has been achieved through the provision of specialised bulk loading terminals, using conveyor belt technologies for ores and grains, and pipelines for bulk liquids. Sometimes oil is transferred to very large or ultra-large crude carriers (VLCCs and ULCCs) by pipeline to offshore buoys, thus avoiding the necessity of docking altogether. However, the net effect of the increased vessel sizes and commodity volumes has been to concentrate the principal flows between a limited number of specialised terminals and facilities in importing and exporting countries. Smaller vessels and/or inland pipelines then distribute the commodities to other areas within the regions concerned.

Fleets of VLCCs and ULCCs expanded rapidly during the 1970s, when the closure of the Suez Canal diverted the bulk of the world's crude oil trade from the Middle East via South Africa's Cape sea route, and the dramatic increase in oil prices rendered the trade extremely profitable. Many smaller and older tankers were scrapped, as were general cargo ships in the wake of the boom in dedicated container ships. The average age of the world's shipping fleets therefore fell. However, for much of the 1980s and early 1990s, the freight shipping industry has been in the doldrums, with excess capacity and low real cargo rates. Significant numbers of bulk carriers, in particular, have been laid up ('mothballed') in the hope of a brighter future. Furthermore, low scrap-metal prices have reduced the incentive to scrap vessels. The average age of fleets has therefore risen substantially (Table 6.1), giving rise to more accidents and very real safety concerns. For example, the number and proportion of foreign ships seized as unseaworthy in British ports over the last few years have risen substantially. In the year to 31 May 1995, the Marine Safety Agency detained 190 foreign ships for compulsory repair, more than 10 per cent of the total inspected. Older and poorly maintained ships, as well as lax standards and inadequate enforcement of regulations, can contribute, along with genuine accidents, to considerable pollution of coastal waters through small, insidious oil discharges and the throwing of rubbish overboard. Such problems are particularly acute in and around major harbours.

A more detailed examination of Table 6.1 reveals that tankers are now on average the oldest category of ships worldwide, almost 17 years as at the end of 1993. Second, there is little variation between groups of

Table 6.1 Age distribution of the world merchant fleet by type of vessel as at 31 December 1993 (percentage of total in terms of dwt)

Country grouping	Type of vessel	Total	0–4 years	5–9 years	10–14 years	15 years and over	Average age (years)[1]	Average age (years) 1992[1]
World total	All ships	100	10.9	18.5	19.7	50.8	15.05	14.91
	Tankers	100	10.4	8.7	17.6	63.3	16.86	16.72
	Bulk carriers	100	11.1	28.2	19.4	41.3	13.61	13.50
	General cargo	100	8.6	17.7	26.0	47.7	15.03	15.04
	Containerships	100	18.0	25.4	17.5	39.0	12.82	12.07
	All others	100	13.0	21.0	21.3	44.8	14.14	13.87
Developed market	All ships	100	10.5	21.1	23.2	45.3	14.44	14.09
economy countries	Tankers	100	6.3	10.6	24.8	58.2	16.65	16.15
	Bulk carriers	100	12.1	31.8	21.1	35.0	12.70	12.73
	General cargo	100	11.6	24.3	28.0	36.2	13.26	13.04
	Containerships	100	19.7	20.4	19.3	40.6	13.07	12.37
	All others	100	14.9	25.1	20.9	39.1	13.17	12.85
Major open-registry	All ships	100	11.2	14.7	16.8	57.3	15.88	15.92
countries[2]	Tankers	100	13.3	5.6	12.4	68.7	17.26	17.36
	Bulk carriers	100	8.4	21.9	18.2	51.5	15.22	15.18
	General cargo	100	9.9	19.9	27.2	43.1	14.34	14.57
	Containerships	100	12.5	28.0	15.9	43.7	13.73	12.70
	All others	100	14.4	19.7	20.3	45.5	14.11	13.98
Subtotal	All ships	100	10.9	17.5	19.7	51.9	15.23	15.09
	Tankers	100	10.3	7.7	17.7	64.2	16.99	16.85
	Bulk carriers	100	10.0	26.2	19.4	44.4	14.13	14.06
	General cargo	100	10.5	21.6	27.5	40.4	13.91	13.97
	Containerships	100	16.7	23.5	17.9	41.8	13.32	12.50
	All others	100	14.7	23.1	20.7	41.6	13.55	13.28
Countries of Central	All ships	100	8.6	20.1	24.0	47.3	14.87	14.75
and Eastern Europe	Tankers	100	8.8	18.8	33.1	39.4	14.13	14.05
	Bulk carriers	100	6.0	27.0	31.7	35.3	13.58	13.34
	General cargo	100	9.0	15.5	17.0	58.6	16.20	16.08
	Containerships	100	8.9	48.9	15.5	26.5	11.29	11.82
	All others	100	12.7	15.9	15.4	56.1	15.56	15.88
Socialist countries	All ships	100	5.1	17.6	16.8	60.5	16.66	16.28
of Asia	Tankers	100	5.1	15.7	12.7	66.6	17.38	16.03
	Bulk carriers	100	4.8	21.2	16.3	57.7	16.23	15.99
	General cargo	100	3.7	11.7	21.1	63.5	17.40	17.37
	Containerships	100	18.9	46.3	7.4	27.4	10.54	9.44
	All others	100	3.2	5.4	15.1	76.3	19.04	19.00
Developing countries	All ships	100	12.2	21.7	19.8	46.4	14.35	14.22
(excluding open	Tankers	100	11.2	11.9	13.9	63.1	16.61	16.65
registry countries)	Bulk carriers	100	17.0	37.8	18.1	27.1	11.12	10.99
	General cargo	100	5.1	11.5	29.2	54.2	16.34	16.39
	Containerships	100	22.8	19.2	17.8	40.2	12.78	12.11
	All others	100	7.4	17.4	28.4	46.9	15.09	14.70

Notes: [1] To calculate average age, it has been assumed that the ages of vessels are distributed evenly between the lower and upper limit of each age group. For the 15-years-and-over group, the mid-point has been assumed to be 22 years
[2] Including Malta and Vanuatu
Source: UNCTAD (1994) *Review of Maritime Transport 1993*, Geneva: UNCTAD

owner countries: the range is only from 17.38 years in the (former) Asian socialist countries to 14.13 in the (former) European socialist states, although the latter is the only group lower than 16.61 years. Worldwide, containerships are the youngest category of vessel, at just under 13 years. Again, there is little variation between groups of countries. The variation is widest for general cargo ships, from 17.4 years in former socialist Asian countries to 13.26 in developed market economies.

Disaggregating the world fleet into groups of owner countries is important in view of the widespread reregistration of vessels by owners in high-income countries seeking to reduce costs by registering them in certain low-income countries where fees and crew costs are lower and the enforcement of regulations on safety laxer. This is also known as reflagging. Ships registered in Europe or North America are subject to stringent controls and trade union requirements to employ local crews at appropriate wage levels. When the same ships are registered in a so-called open-registry country, of which Panama, Liberia, Cyprus, the Bahamas and Bermuda are the world's largest, these restrictions do not apply. Typically, only the most senior officers are then employed from the 'home' country, while cheaper foreign labour is recruited for less skilled jobs. Many thousands of Greeks, Filipinos, Indians, Pakistanis and Sri Lankans, in particular, are now employed in this way, although significant numbers of other nationalities also crew on ships belonging to other countries.

Tables 6.2 and 6.3 indicate just how important open-registry countries are in terms of all categories of vessel. In contrast to the very high level of local ownership in the Danish and Norwegian international registries, the five open-registry countries own negligible shares of the tonnages registered there. Indeed, in three of them the figure is zero and in the Bahamas it is 0.4 per cent. Only Cyprus has some local ownership (Table 6.3). Greece, Japan, the USA, Hong Kong, Norway, the UK and Germany are the principal true owners of vessels in open-registry fleets, in terms both of tonnage and number of ships. However, South Korea, China, Denmark, Russia, Taiwan and Singapore are also substantial owners.

By contrast, most poorer countries of the South own few, if any, ocean-going ships. They are therefore dependent on foreign vessels. Moreover, although overseas trade, especially with Europe and North America, forms the bulk of international trade for most such countries, the volumes concerned represent a low proportion of total global trade.

Table 6.2 Tonnage distribution of major open-registry fleets[1] as at 31 December 1993

	Oil tankers		Dry bulk carriers		General cargo		Containerships		Others		1993 total		1992 total	
	Ships	Thousand dwt	Ships	Thousand dwt	Ships	Thousand dwt	Ships	Thousand dwt	Ships	Thousand dwt	Ships	Thousand dwt	Ships	Thousand dwt
Liberia	394	49,030	399	25,263	283	4,971	104	3,039	281	6,051	1,461	88,354	1,508	91,757
Panama	341	32,857	576	26,806	1,523	14,706	183	4,460	528	4,163	3,151	82,992	2,927	73,524
Cyprus	81	6,168	495	19,708	532	5,339	57	871	65	583	1,230	32,669	1,168	30,384
Bahamas	161	17,913	147	8,151	368	4,678	32	725	213	1,595	921	33,062	896	31,874
Bermuda	18	3,755	8	247	14	111	5	112	33	873	78	5,098	76	5,467
Total	995	109,723	1,625	80,175	2,720	29,805	381	9,207	1,120	13,265	6,841	242,175	6,575	233,006

Note: [1] Ships of 1,000 grt and above
Source: UNCTAD (1994) *Review of Maritime Transport 1993*, Geneva: UNCTAD

Table 6.3 Tonnage owned by the nationals of, and registered in, the country of registry in the total fleet of the most important open and international registers (thousands dwt[1] as at 31 December 1993)

Country of registry or register	Total tonnage registered in the country of register	Tonnage owned by nationals of, and registered in, the country of registry	Share of tonnage owned by nationals in the total registered fleet (%)
Liberia	88,353	0	0.0
Panama	82,992	0	0.0
Cyprus	35,673	3,003	8.4
Bahamas	33,190	128	0.4
Norwegian International Ship Registry	33,138	30,663	92.5
Danish International Ship Registry	6,389	6,358	99.5
Bermuda	5,098	0	0.0

Note: [1] Ships of 1,000 grt and above
Source: UNCTAD (1994) *Review of Maritime Transport 1993*, Geneva: UNCTAD

Freight rates are frequently therefore higher than for equivalent distances and consignments on high-volume routes. Transit times are also likely to be slower, because of non-direct shipping services, transshipment, older vessels and loading technologies and lower priorities attached to such flows. Port efficiency and levels of maintenance may also be suboptimal. Considerable effort and investment have therefore been expended in recent years on improving or rehabilitating facilities, management and regulatory environments in ports and on other related infrastructure. For example, the period 1979–88 was designated as the United Nations Transport and Communications Decade in Africa (UNTACDA). Similarly, the international transport sector (especially ports and railways) absorbed over half the total expenditure of the Southern African Development Co-ordinating Conference (SADCC), restyled the Southern African Development Community (SADC) in 1992, as the focus for its collective efforts to reduce dependence on South African routes and trade during the apartheid years of the 1980s and early 1990s.

Land transport

Technological change, especially the ongoing trend towards larger and more specialised heavy goods vehicles (HGVs) (also known as lorries or trucks), has also affected long-distance land transport. The extent of control over maximum permissible vehicle mass varies between countries, but opposition to increases is usually far less – if indeed it even exists – than in countries of the North. Although in theory the introduction of larger vehicles reduces the number required to carry a given tonnage of freight, in practice the volume of freight moved by road has been growing rapidly almost everywhere. This reflects a combination of newly generated traffic and the diversion of increasing proportions of freight from rail to road. Road has long held the edge (i.e. comparative advantage in neoclassical economic terms) over rail for short distances, but railways have been struggling to retain their traditional bulk cargoes over longer and longer distances in recent decades.

The factors behind this shift represent a combination of:

- larger and better HGVs, enabling lower rates per tonne to be charged while shifting larger consignments;
- trunk road improvements, facilitating quicker road journeys with reduced wear and tear on HGVs and damage to loads;
- avoiding the need for transshipment and handling costs, providing a quicker and direct door-to-door service;
- changing regulatory conditions which have often liberalised the road transport sector, enabling more competitive haulage policies and reducing rates, while state-owned railways remain heavily regulated and bureaucratic, perhaps even retaining their traditional 'common carrier' obligation to carry any goods at tariffs often too low to cover the marginal costs;
- cutbacks in the rail networks and services on remaining lines as a cost-cutting measure to increase efficiency and reduce operating losses. This reduces further the attractiveness of rail to any traffics not served directly and increases the range of routes for which direct road transport becomes attractive.

However, the often dramatic increases in road transport – of both freight and passengers – have generally outstripped the ability to extend or upgrade the infrastructure appropriately, or even to maintain the existing roads. Congestion on main trunk routes is now common, and the toll of accidents is rising almost everywhere, despite efforts to educate drivers about the dangers of speeding, reckless overtaking, overloading and poor maintenance. Studies in a wide range of countries have found that a high proportion of HGVs are overloaded, thus increasing the risk of damage to the vehicles and certainly causing damage to road surfaces. Many HGVs are old and in poor condition, generating quite heavy localised air pollution with their thick diesel emissions. If road maintenance is inadequate, corrugations, potholes and crumbling of the tarred surface can result. Particularly in areas where heat and intense seasonal rainfall are the norm, damage can be dramatic and occur very rapidly.

Such problems are often exacerbated where heavy domestic traffic is supplemented by substantial numbers of transit vehicles to landlocked countries. This is well exemplified by the cost to Kenya of large numbers of HGVs carrying imports and exports to Uganda, Rwanda and Burundi right across the country from the port of Mombasa to the Ugandan border. The most heavily used stretches, namely those from Mombasa to Nairobi, and from Nairobi to Nakuru, have been deteriorating rapidly over recent years, despite the resurfacing of the Nairobi–Nakuru section. Transit fees and tolls did not cover the so-called track costs, i.e. real costs of road maintenance, until they were increased substantially in 1994/5. Moreover, a pipeline has been constructed to carry petrol from Mombasa to Eldoret, the next major town after Nakuru. Transit petrol tankers for neighbouring states are now required to load up at the Eldoret terminal instead of in Mombasa. By early 1995, shortly after it began operations, the removal of the substantial numbers of transit tankers from the busiest sections of Kenya's main trunk road (Plate 6.7) had already improved conditions markedly. Nevertheless, industry sources are complaining that they pay more for petrol at the Eldoret terminal than it would cost them to purchase the load in Mombasa and haul it there by road or rail.

Long-distance and intercity passenger transport by road is also increasing, for similar reasons to those listed above with respect to freight. Passenger rail services are extremely slow and are losing customers and suffering service cuts almost everywhere. Although in most low-income countries, long-distance bus or shared-taxi travel is often crowded and uncomfortable in old vehicles, private and public operators in some countries such as Cameroon and South Africa have introduced higher quality buses at competitive fares, which are proving attractive to air as well as rail passengers. More generally, though, as discussed in relation to urban bus and rail services in Chapter 5, shortages of imported spare parts and other problems often keep vehicles off the road.

Few countries of the South can afford major infrastructural investments such as the rehabilitation or construction of new trunk roads themselves; these are therefore commonly funded at least partially by

Plate 6.7 Until the recent opening of the petrol pipeline from Mombasa to Eldoret, Kenya's main trunk route was congested by large numbers of transit tankers *en route* to Uganda, Rwanda and Burundi

means of foreign aid loans, grants and technical assistance. However, these projects seldom include a maintenance component, so that in countries where the resources or organisational capacity for such tasks do not exist, newly completed roads soon deteriorate, undermining the value of the initial project.

Conclusions

The focus of this chapter has been on longhaul transport by air, sea and land, indicating not only how the respective modes have developed over recent years, but discussing the impact of technological change for those modes themselves and for the development of the cities, regions and countries concerned. There are clearly winners and losers in all such contexts, but innovations may well have wider developmental impacts which are quite strongly positive on balance.

A case in point is the growth of civil aviation and the role of wide-bodied jetliners in promoting mass tourism, now one of the principal generators of foreign exchange in many Third World countries. The extent of the developmental benefit depends on the market share held by national as opposed to foreign airlines in the absence of a revenue-sharing agreement, the import content of the industry, the level and nature of local employment generation, and the social and environmental impacts of tourism.

In terms of water transport, bulk cargoes and containerisation have speeded up transit times and cut the labour cost of freight traffic, but have also concentrated traffic increasingly into a small number of large hub ports, where consolidation and transshipment for local or regional distribution and collection occur. Small ports with inadequate facilities and traffic volumes, as well as the people employed in stevedoring and maintaining traditional ships, have certainly lost out. The costs of the new technologies and the facilities required to use them have been considerable, and most poor countries have had little choice in whether to adopt them.

Rail passenger and freight services have suffered increasingly from road competition over recent decades, as a result of larger, more comfortable vehicles, more direct and reliable services, and more favourable regulatory regimes. Rail services are having to concentrate on bulk and specialised services. Such changes have also greatly stimulated the development of inter- and multimodal transport involving road or rail as well as water and air modes. Quick, reliable door-to-door service forms the basis of competition for both long-distance domestic and international traffic, although travel for many poor passengers remains slow, uncomfortable and often not reliable.

Key ideas

1 Technological changes have opened many new development opportunities, although not without costs, while poorer countries may have little effective choice over whether or not to adopt them.
2 Civil air transport has been growing very rapidly, stimulating the expansion of dedicated tourist development in many 'exotic' tropical locations and also the air freight of relatively high-value fresh cut flowers, soft fruit and vegetables to markets in the North.
3 While containerisation and the evolution of bulk commodity shipping have adversely affected small coastal and riverine ports and their immediate hinterlands, major ports have experienced rapid

development, sometimes forming integral parts of extended metropolitan regions.

4 Road transport has been progressively eroding traditional rail traffics in both passengers and freight. It tends to be quicker, cheaper and more direct.

5 Intermodalism has also increased, offering fast, reliable and competitively priced services.

7
Transport and development: policy and planning into the twenty-first century

Introduction

Each of the preceding chapters has addressed a particular set of issues, outlining the current situation and recent trends, and teasing out the intricate relationships between transport *per se* and the wider development contexts in which it is embedded. Different theoretical frameworks have been sketched and evaluated with reference to relevant processes and empirical data from a range of countries and continents. Since the strands of each chapter have been drawn together in a separate concluding section, complete with a set of key ideas, this material will not be repeated here. Instead, some of the principal arguments running through the book will be drawn together by means of a forward look towards the end of the millennium and the dawn of the twenty-first century. This is not intended as an attempt to predict long-term future trends but more as an outline of some key transport policy and planning issues which will need to be addressed in the light of how current situations are likely to unfold, if the development aspirations of countless people in the countries of the South are to be even partially met.

The evidence presented in this book suggests clearly that no single theory or analytical framework can adequately account for the diversity of conditions and processes across the South, let alone the nature of relationships between different regional blocs or between South and North. Equally, no theoretical perspective can be *entirely* discarded, even if now widely discredited as a claimant to be 'the best' or 'the only credible' development theory. It is also clear that conditions in the South are becoming increasingly heterogeneous, and this has been amply illustrated with respect to the transport sector and how it relates to wider development processes in specific contexts. It is against this background that the so-called postmodern turn in development theory should be seen. Provided that this is taken to mean that discordant voices, different explanations and divergent paths can have simultaneous legitimacy, then it certainly represents an important step forward. However, any meaningful social organisation or action, let alone policy and planning, still requires some accommodation, some sense of collective rationale and will.

Additionally, both theoretically and in terms of appropriate policy interventions, a holistic perspective remains as important as ever. This should not be rigid and prescriptive but a flexible blend of the macro- and microscales, of what is commonly called top-down and bottom-up involvement in planning processes. Local communities and organisations should certainly be centrally involved throughout, having shared responsibility for and control over the relevant resources and outcomes. After all, they are likely to have a more intimate knowledge of local conditions and social relations than outside facilitators, 'experts' and politicians. However, their perspectives are not the only valid ones: they frequently do not know the wider picture or have the insights or experience to negotiate the wider social, political and economic contexts and processes which impact directly upon their situations.

Appropriate theoretical frameworks for the future should avoid the narrowly prescriptive and, as discussed especially in Chapter 3, be based on seeking the limits to generalisation on the basis of underlying forces and processes. This provides the wider understanding which can be adapted to local contexts. Conventional urban and transport planning have conspicuously failed in this respect. The dominant ethos remains very technicist and formalistic on the engineering and town planning sides, and narrowly economistic in terms of transport economics. As this book has demonstrated repeatedly, the social, political and environmental dimensions are as important to appropriate and progressive interventions as 'whether the sums add up' in terms of partial cost–benefit analyses or engineering feasibility studies. Unless they fit the context and are accepted by the people supposed to use and benefit from projects, the most brilliant engineering design or economically optimal solution will have been in vain.

Liberalisation and privatisation strategies

As mentioned in several previous chapters, one of the dominant international economic policy trends affecting countries of the South since
the early 1980s has been the promotion of liberalisation and privatisation policies. They lie at the heart of structural adjustment and economic
recovery programmes, the adoption of which has been achieved through
vigorous political pressure by the Northern donor community as part of
the so-called 'economic conditionalities' for continued aid from both
bilateral and multilateral sources. Such structural adjustment lending has
been provided in stages, with the real threat of suspension if recipient
countries do not demonstrate sufficient commitment and progress.

Liberalisation, or deregulation, as it is also known, refers to the
relaxation or removal of regulations and restrictions on economic
activity imposed by the state. Privatisation is the transfer of ownership
of economic enterprises from the state to the private sector. The claimed
rationale for these policies is that experience in a wide range of
countries has shown state ownership and operation to be inefficient,
bureaucratic and prone to corruption and that the private sector would
do a better job. The philosophy underlying such policies is that of
supposedly free market economics, which argues that the direct role
of the state in economic activity should be limited to the provision of
public goods for which externalities preclude private supply, and the
setting and enforcement of the minimum of regulations necessary to
ensure quality and fair competition in private enterprise.

Experience to date

There is certainly much evidence to support the claims regarding state
involvement in the transport sector, some of which was discussed with
respect to buses in Chapter 5. Parastatal transport corporations have
often absorbed disproportionately high investment funds and cash injec-
tions to cover operating losses but, especially in countries hit by debt
crises and having to implement structural adjustment programmes, cash
starvation has become common. However, private sector inefficiency
and corruption are also well known. There are examples of efficient and
effective public transport and of public–private partnership. Furthermore, many transport services, e.g. MRT and LRT systems, are inher-
ently monopolistic in nature and there is no evidence that private
monopolies are necessarily more efficient than public monopolies or

any reason why they should be. Indeed, the latter may well be socially preferable.

It is important to emphasise that liberalisation and privatisation policies are not being implemented solely in the South. In fact, they originated in the North and have been pursued vigorously there too by an increasing number of governments of all political persuasions. Many of the experiences are parallel, despite the often substantial differences in the wider political economy. In any case, the history of transport policies over the last century or so shows successive phases of regulation and deregulation and of periodic shifts between private and public ownership in many countries. Both during colonial rule and since independence, countries in the South have tended to follow such trends in the North, especially in their (former) colonial power, albeit often with a lag of a few years.

Four general objections to the thrust of current policies can be raised, however:

- Adequate account has not been taken of such politico-economic and social differences and the policies have often been promoted in simplistic, blueprint fashion which reflects ideological commitment more than local conditions in any specific case.
- If the objective is the promotion of greater efficiency and effectiveness (in this case of transport), liberalisation and privatisation should be considered separately. Too often they are seen as inextricably linked, sometimes even to the point of being regarded (incorrectly) as synonymous or as two sides of the same coin.
- Greater flexibility should be shown in determining the degree of such policies appropriate in any given context. Far too often, the assumption has been that total deregulation and privatisation are optimal. Particularly in Third World countries, partial implementation may actually provide greater overall benefits and promote more balanced development. In addition (and not only in the South), the substantial scope for promoting patronage and self-enrichment through sell-offs of public enterprises has generally been overlooked.
- Even though some countries in the South have adopted such policies willingly or in an effort to pre-empt the suspension of foreign aid, the way in which donors have imposed their 'conditionalities' in most cases has been resented in the South as a hostile act representing a form of economic imperialism; as motivated primarily by self-interest because Northern firms have much to gain from the opportunities thus

provided; as disrespectful and indicative of Northern arrogance; and/ or as undermining national sovereignty in policy making.

Future directions

There is little prospect of any significant change to the basic thrust of liberalisation and privatisation policies in the international arena in the foreseeable future. Certainly, the end of socialist and hard nationalist experiments at both extremes of the political spectrum has meant that hardly any countries still adhere to rigid and doctrinaire views about the absolute necessity of state ownership and control of their economies. Indeed, some former Marxist-Leninist states like Ethiopia, Mozambique and Vietnam have thrown their doors open to international capital in the last few years; the first two are permitting virtually unconstrained private enterprise, with plans to privatise most if not all parastatal corporations, while Vietnam has introduced a homegrown version of economic liberalisation known as 'Doi Moi' and is utilising large foreign investment to fuel an increasingly aggressive export-oriented industrialisation programme. Even Fidel Castro's government in Cuba, so long a bastion of Marxist defiance, is quietly permitting some domestic liberalisation and promoting foreign tourism as a source of desperately needed foreign exchange. There are also signs that North Korea, now ruled by Kim Jong Il, son of the late Kim Il Sung, may be ready to move away from its self-imposed and rigidly enforced autarchy (international isolation).

Liberalisation

It is evident that regimes of all political persuasions have experienced similar problems with state-owned transport and – often justifiable – claims regarding unfair treatment of public operations compared with the private ones with which they are in competition. In practice, some measures favour public transport but others are advantageous to the private sector. Those favourable to the state sector include the rebates or subsidies given to parastatals, railways and other publicly owned transport enterprises in respect of fuel, spare parts and maintenance; and the availability of credit or investment loans below commercial rates of interest. Those advantaging private operators include the common carrier obligation still applicable to many national railways and state road haulage firms which requires them to accept unprofitable freight (largely because the tariffs are too low to cover marginal costs); the statutory obligation of some railways to meet their total costs (including capital costs); and the fact that in many countries the track costs (i.e. cost of damage to roads and contribution to replacement) of commercial road vehicles are not fully covered by licence fees and toll charges.

It is arguably desirable to continue with and improve the targeting of deregulation policies, perhaps in conjunction with some form of privatisation if appropriate. However, the frequently blunt way in which such policies have been formulated and implemented in widely differing situations should be refined. They need to be based upon adequate local research and meaningful in-country consultation. One of the key objectives should be to 'level the playing field', allowing competition between public and private sector firms on a fair and equitable basis. This certainly requires the remedying of often hidden subsidies of the type cited in the previous paragraph. It also means enabling the management of parastatal enterprises and public sector transport enterprises to operate in more flexible and commercially oriented ways, e.g. by removing common carrier obligations, permitting managers to amend passenger fares and freight rates and to negotiate access to commercial credit subject to necessary guarantees and safeguards.

However, some conflict between pure commercial and national considerations is often unavoidable. Even though the managements of many national railways have gained greater autonomy and flexibility, governments are generally keen to reserve certain powers of approval for themselves, e.g. on the abandonment or introduction of new services and the closure of existing branch lines. For example, Table 7.1 highlights the division of powers between the management, Board of Directors, and Council of Ministers representing the governments of Tanzania and Zambia in respect of the jointly owned and operated TAZARA Railway. While fare changes, salary adjustments and the granting of concessions and rebates of up to 10 per cent can be taken by management, more substantial changes require Board approval, while substantive policy matters, the raising of capital and appropriation of any surplus are powers reserved by the Council of Ministers. Clearly, as a joint venture between the two countries, questions of the respective governments' interests have to be balanced; while this may be a handicap in certain respects, it reduces the direct control which a government could exercise over an operation solely within its borders.

Table 7.1 Distribution of decision-making powers for the TAZARA Railway

Functions	Powers of Council of Ministers	Powers of Board	Powers of management
Any important question on railway policy	Reserved	None	None
Changes in tariff rates, fares and other charges	None	Changes exceeding 10%	Changes up to 10%
Tariff adjustments upon exchange rate changes in Zambia and Tanzania	None	Full powers	Full powers
Fixing rates and fares which are not specified in the Tariff Book	None	None	All items
Grant of concessionary rates or rebates	None	Concessions of 25–50% on any item	Concessions of 1–10% on selected list of items
Raising of new capital	Reserved	None	None
Approval of appropriation of surplus	Reserved	None	None
Approval/acceptance of sale, disposal or write-off of property	None	Items valued above US$2,000	Items valued less than US$2,000
Revision of salaries, wages or allowances	None	Changes over 10% for all scales	Any change up to 10% for scales 1–6

Source: Mwase, N. (1994) 'The liberalization and deregulation of the transport sector in sub-Saharan Africa', *African Development Review* 5(2): 74–87

Privatisation

Unbridled competition is often no more desirable or efficient than total regulation. If 'artificial' quantitative restrictions, e.g. haulage licence quotas, are removed too rapidly, the industry can easily be thrown into turmoil as numerous new operators seek part of the traffic in question. Partial or phased relaxation has often proved more helpful in this respect. Similarly, relaxation of quantitative restrictions increases the argument for qualitative control systems to ensure that appropriate codes of conduct and safety are adhered to. A common tactic in the absence of such systems is for certain hauliers to undercut their competition by reducing their rates to a level which no longer covers their marginal and average costs. Such 'predatory pricing' reduces their ability to pay staff and maintain or replace vehicles adequately, and has been associated with increased accident risks.

In a similar vein, full and rapid privatisation may simply replace a public monopoly with a private equivalent, lacking any pretence at social obligations and with more freedom to increase prices. In practice, full privatisation often also means pumping additional public resources into loss-making operations in order to restructure and render them attractive to private investors, selling them at discounted prices, and leaving only the least attractive operations in public hands. For these reasons, in a number of countries, joint private–public ventures are proving more successful and less politically contentious than outright privatisation. There is also a strong case for innovative policies to increase productivity and efficiency by involving staff more directly in decision-making, responsibility and even as shareholders rather than selling all the shares on the open market and to foreign investors. This could apply to private as much as to public companies.

A current example of wholesale privatisation is the plan of the new Mozambican government to sell off the national airline, LAM, as part of the neoliberal economic programme that has been implemented progressively during the 1990s. Already one of its two leased Boeing 737 aircraft has been returned to the owners; and services have been cut. It is clear that the country cannot afford the necessary new investments to modernise equipment and the airline is possibly too small to become sustainably profitable in the medium to long term. However, only very limited opposition to the proposed sell-off has been aired by powerful voices within political circles to date, despite the apparent lack of serious consideration that appears to have been given to alternative scenarios such as a joint venture or joining the new regional airline, Alliance Airline Services (AAS), launched by South African Airways and the Tanzanian and Ugandan governments on 30 June 1995.

Finally, it is worth noting that substantial regional differences may exist in terms of historical situations and current policy debates. For example, Southeast Asia has a strong tradition of private ownership in urban transit, especially of bus companies and paratransit operations. These are often efficient and provide good service quality. Sometimes these operate in parallel or in competition with public bus services. The latter, however, frequently operate at a loss and require substantial subsidies. Paradoxically, states in the region show considerable interest in increasing public ownership – a policy which distinguishes transport from other economic sectors there – on grounds of increased productivity and improved overall service, alleviating traffic congestion, reducing air pollution, and promoting integrated planning and the like. This is certainly at odds with current trends in Africa and much of Latin America.

Intermodal and integrated transport

Intermodalism

From the evidence presented in Chapter 6, it seems clear that the importance of intermodal transport, involving the use of several modes in providing integrated, door-to-door services, will continue to increase. Currently, such services are provided both domestically and internationally by a range of public and private specialist operators, covering everything from express document courier services to parcels and dedicated trainload consignments of particular commodities. They are far more widespread in the advanced industrial economies of the North, but are increasingly available to and from major centres in the South as well. Once again, though, there is frequently a technological bias in that document and parcel courier services operate through major international airports. Hence, it is straightforward nowadays to consign a package for delivery within 24 or 36 hours between Europe or North America and, for example, Buenos Aires, Rio de Janeiro, Abidjan, Johannesburg, Manila or Seoul. Particularly in countries with less well-developed internal infrastructure, delivery outside large cities may take far longer or not even be available.

On the other hand, dedicated intercontinental traffics of high-value export commodities from the South do now often operate on specific routes from areas of cultivation or manufacture which may be several hundred kilometres from the gateway airport through which they are exported. For example, cut flowers are grown on a large scale around Lake Naivasha and in the Kericho area of western Kenya. They are picked and packed under rigorously maintained temperature-controlled conditions to maximise longevity and minimise damage. Then they are transported by road to Nairobi's Jomo Kenyatta International Airport for transfer to aircraft destined mainly for the Netherlands but also the UK, Germany and South Africa. They are then auctioned and distributed to florists and other shops within hours of arrival. As new export opportunities, facilitated by technological change and improvements in infrastructure, become available, such commodity flows and the integrated services necessary to ensure swift, reliable and safe delivery will expand. These are likely to be of both the dedicated variety illustrated above and general services which consolidate numerous small consignments for different clients along the lines of couriers or 'groupage' by freight hauliers.

Other forms of intermodal service involve the shipment of containers intercontinentally using combinations of road, rail and sea transport. Such services are also constantly increasing and embracing new routes, in accordance with levels of demand, the regularity of the flows, the quality and reliability of the various transport sectors and client requirements in terms of total transit times. For example, some shippers in the Far East and elsewhere in the Asian Pacific Rim now find it more efficient to ship TEUs to the West Coast of the USA for rail haulage across the continent to an East Coast port for onward shipping to Europe, rather than undertaking the entire voyage by sea. Increasingly also, such operations are undertaken on a hub-and-spoke system, involving the use of a very small number of gateway hubs, perhaps only one or two per region. Transshipment and onward haulage by coaster, rail or road as appropriate take place to and from these primary nodes. Similarly, within individual countries in both the North and South, TEUs are now commonly transported by road, even where railway services do exist, because of speed, reliability and sometimes also cost advantages.

All such trends will intensify, mediated by changing relative advantages of particular routes and modes of transport in terms of reducing transit and transshipment times and losses *en route*. What is important to bear in mind is that these changes frequently have development impacts both at the level of the individual port or transshipment point which either gains or loses the traffic and at the macro-level, where the time–space geography of the country or region is changing. In other words, integration and disintegration are constantly occurring differentially across national, regional and even larger territories. Perhaps the most dramatic current example of this is the way in which the rapidly expanding computer networks linked to the Internet are redefining telecommunications (and to some extent replacing a need for personal travel) across the globe. Again, the North generally has far more equitable and widespread access than do most countries of the South, but it is important to underline that access and lack of access may be separated by extremely short physical distances – perhaps even from one house or neighbouring settlement to the next.

Integrated transport

The need for greater co-ordination and integration in transport services can be considered at two distinct scales, namely the urban and the international. Each will be addressed only briefly here.

The urban scale

Integrated urban or metropolitan transport is being promoted in a number of megacities and rapidly growing urban regions on all continents. Some of the best documented efforts have been in Santiago, Porto Alegre and São Paulo in Latin America and in Hong Kong and Singapore in Southeast Asia. This involves planning for all modes of transport – as discussed in Chapter 5 – on a coherent and integrated basis, with the provision of interchanges between them at MRT, LRT or suburban rail stations, for example. The logic here is that piecemeal approaches, addressing only a single transport mode or the needs of a particular locality within a metropolitan area, cannot ease the overall problems and may actually increase energy consumption while exacerbating environmental problems, the difficulties of travel across the metropolis and the low accessibility experienced by the urban poor. Integrated urban transport planning and management do not consist entirely of logistics and scheduling; traffic management and restraint, along with road safety issues, are also integral components.

However, whatever the rhetoric and intentions, successful integrated metropolitan transport systems remain very much the exception rather than the rule. The traffic chaos of Bangkok and Jakarta is rather more common, demonstrating well the need for a more holistic approach. The precise nature of policies to be adopted will depend on the existing transport systems, the quality and economic viability of services, the gap between supply and demand for urban travel, the physical character of the urban area concerned, attitudes to public versus private ownership and operation, attitudes to paratransit and traditional forms of non-motorised transport if they exist, economic conditions in the country and metropolitan government concerned, and, very importantly, the extent of political commitment to tackling the problems and seeing the approved plans implemented. In general, the objective should be to enhance what already exists, optimising the potential contribution of each existing and planned mode of transport so that the whole represents more than the sum of the individual parts.

For example, paratransit should be seen as making a potentially positive contribution, rather than representing an unwelcome residual of a traditional past that ought to be eliminated. As concerns for energy efficiency, urban congestion and pollution increase, the potential contribution of non-motorised and certain forms of motorised paratransit should be more readily appreciated. Nevertheless, paratransits may not be appropriate for all parts of, or regarded as important by all communities within, a given urban area. Just as MRTs or LRTs can play a major role along high-volume, longer distance routes, especially if integrated with bus routes and feeder services, so many forms of paratransit are well suited to densely populated areas, especially where a substantial proportion of the inhabitants are on low incomes, where roads are narrow and irregular, or in (semi-)pedestrianised areas from which motor vehicles are excluded.

The international scale

There is growing momentum around the world towards the formation of regional economic groupings or more clearly defined blocs designed to promote greater intraregional trade and co-operation and/or to enhance the strength of a given region in its collective trading relations with other regions. Many such initiatives, often modelled on the European Union or Council for Mutual Economic Assistance, have failed in the past on account of overambitious objectives, the inability of member countries to support considerable bureaucratic structures, inadequate economic complementarities between member states and the reluctance to trade off national self-interest against collective benefit.

The current efforts at regional integration have diverse origins, objectives, internal coherence, ways of operating and track records. However, many of them are realising that if member states are to increase trade and other forms of interaction with one another then better and more reliable means of transport linking them are necessary. Often, as explained in Chapter 3, individual Third World countries retain, at least in part, a strong colonial legacy of rail and principal road networks which traverse their territory to the principal port(s) on the coast rather than providing internal accessibility or integration. This is a real hindrance to the objectives of greater regional collaboration, and is being reflected in a number of substantial projects, funded largely by the donor community, to rectify the shortcoming or upgrade existing links, e.g. the Pan American Highway, Trans-Amazonian Highway, Trans-Africa Highway and the Trans-Sahara Highway. The increasing cost of airline operations and the lack of long-term profitability of many small national carriers is also giving new impetus to the formation of regional airlines in the tradition of the defunct East African Airways and the ailing Air Afrique. One new example was referred to in the previous section, while the Southern African Development Community (SADC) is reportedly considering a truly regional airline.

Even where international links are already good, as in the Singapore Extended Metropolitan Region discussed in Chapter 6 or the Mexico–USA border zone, the intensification of interaction and trade as a result of the integrated manufacturing processes is prompting the development of higher capacity, regular and reliable transport links, such as the proposed new causeway linking Singapore with Johor. These trends will continue and intensify, enabling functional metropolitan areas to expand both territorially and in terms of the proportion of populations and value-added they attract. There is certainly scope for successful experiments in the South to be 'exported' to metropolitan areas in the North, thus reversing the well-established flow.

Safety and sustainability

These two issues will undoubtedly form more central components of transport and traffic policies over the coming years. Whereas vehicle and passenger safety have long been a major concern in the North, they have tended – in practice if not rhetoric – to be neglected in most low-income and lower-middle-income countries. Governments and the population have often regarded them as of secondary importance and a poor safety record as being an unfortunate (and perhaps inevitable) side effect of poverty and the push for economic development. Much the same has applied to environmental conservation and concerns to promote more sustainable development: poor people engaged in a daily struggle for subsistence and governments seeking rapid modernisation have frequently regarded campaigners and international pressure to adopt different policies as an irritant or perhaps as totally irrelevant in that any long-term programmes presuppose that short-term survival is assured.

The challenge of sustainability

Change will not be rapid or uniform; however, there is mounting evidence that the urgency of the situation is generating increasing pressure for action from within individual countries as well as at an international level. The reports of two international commissions, the World Commission on Environment and Development (the Brundtland Commission) and the South Commission (the Nyerere Commission), published in 1987 and 1990 respectively, were widely welcomed as breaking new ground. In their wake, global conferences such as the UN Conference on Environment and Development, held in Rio de Janeiro in 1992, the UN Population and Development Conference in Cairo in 1994, and the UN Social Development Summit in Copenhagen in 1995 are providing international momentum and commitments, however diluted by the inevitable negotiations behind the scenes, to which government leaders are subscribing. The agreements signed contain timetables and targets, and seek to reduce the gap between North and South in terms of the standards which apply, the levels of consumption of energy and production of greenhouse gases, and different perceptions of what constitute the principal problems and acceptable ways of addressing them.

In respect of transport, the most important sustainability issues relate to consumption of non-renewable energy, which is derived overwhelmingly from oil, the growing contribution of vehicle exhaust emissions to problematic and even health-threatening levels of atmospheric pollution in cities, the rising problem of traffic congestion which now threatens the efficiency and effectiveness of urban transport over lengthening periods of the working day, and the growing use of non-recyclable materials in vehicle construction. Waterborne and air transport have their own specific problems of pollution, especially where traffic is dense, as in and around major ports and navigation channels. The economic and financial sustainability of certain current transport modes and systems are also in question, for reasons discussed above and in Chapters 5 and 6.

Improving safety

A growing body of research has highlighted the terrible human cost of road accidents in Third World countries. This results from a combination of poor vehicle maintenance, inadequate or deficient road infrastructure and safety engineering, inadequate driver training and skill, and reckless behaviour by drivers and pedestrians. The relative importance of these factors varies between localities and each can best be addressed in different ways according to local circumstances.

Common approaches to improving the condition of vehicles include introducing a system of annual inspections as a prerequisite for licensing, higher spot-fines for major deficiencies observed on the road, and the impounding of persistently offending vehicles. Vehicle emission tests can be included in such procedures, although the low or variable quality of fuel may render this problematic. It could also be made compulsory for vehicles to carry adequate insurance cover. Naturally, all the measures just referred to are far more difficult to enforce in rural areas; besides, where roads are in poor condition, they constitute a major source of wear and tear on vehicles beyond the control of drivers or vehicle fleet operators.

Many countries now have infrastructure improvement programmes, with annual targets for the upgrading of poor gravel roads, tarring of major routes, maintenance and the installation of traffic lights and other management measures in urban areas. These are often reliant on foreign aid and sometimes also technical assistance. As explained in Chapter 4, rural trunk or feeder road schemes may form part of integrated rural development programmes, providing wider benefits to the surrounding areas. However, such expectations are not always met, while construction and maintenance programmes frequently lag behind schedule as a result of shortages of funds, skills, equipment or spares as well as political problems. Countries everywhere are now increasingly conscious of the need to promote maintenance as well as new construction and that, in the face of budgetary or other constraints, maintenance usually represents a far more effective use of limited funds than new projects. The consequences of a lack of maintenance are now widely appreciated (see Chapter 4 and Plate 4.2).

Pedestrian behaviour can be influenced by public education programmes, targeted especially at schools and at vulnerable groups such as the elderly and those with impaired sight or hearing. Again, however, the provision of adequate pedestrian crossings and other measures to segregate them from heavy vehicular traffic and particularly dangerous intersections need to be prioritised by urban transport departments. Driver skills and behaviour should be improved primarily through more rigorous training and testing, with periodic refresher courses for commercial and public transport drivers who earn their livings on the road. Supportive measures would include heavier penalties for unlicensed driving, making it a legal requirement that licences be carried when driving, and the introduction of drivers' licences which include a photograph of the holder and are more difficult to forge. The technology for this is now more widely and cheaply available. Countries which have introduced and enforced legislation limiting the length of bus and lorry drivers' shifts and imposing minimum rest periods have generally experienced a reduction in fatigue-related accidents. Although enforcement would be difficult in many poor countries, such measures are also worth considering seriously.

Road accidents claim more lives than all other modes of transport together in most countries, as a result of numerous tragedies claiming small numbers of lives each, except where buses or unusual multiple vehicle pile-ups are involved. Apart from the overall cost in terms of lives, medical treatment, damage and repairs, among the dead are many religious, political and business leaders and other professionals for whom long-distance work-related travel is regular. Developing countries can ill afford to lose so many skilled people.

Rail, shipping and aviation disasters are less frequent but more dramatic in terms of the number of people killed and injured each time. Consequently, they tend to be more widely publicised. Aviation safety is generally relatively well maintained and policed, although there is currently concern being voiced about an illicit international trade in non-genuine and second-hand spares, some of them worn out or damaged in accidents. Such cheaper parts have found their way into use with leading international carriers and have been blamed for causing one or two recent crashes. The potential impact of widespread use by cash-starved Third World airlines using older aircraft could be serious.

Rail and shipping safety standards and maintenance levels tend to be very variable, an issue discussed in Chapter 6 with respect to the widespread use of open registers as a cost-cutting measure. Stricter enforcement, coupled with better crew training and certification, is necessary in some countries. Although most attention tends to be devoted to ocean-going vessels, the largest number of accidents and casualties actually occur in short-sea, river and lake crossings. Countries comprising island archipelagos or where interisland transport is important, like the Philippines, Indonesia and the Pacific Island and Caribbean states, rely more on ferry transport than most others and are therefore more vulnerable. Reports of ferries capsizing in bad weather, through overloading or mechanical breakdown, are too frequent, and casualties are usually high. Inland waterways are often hazardous too: on 11 and 12 April 1995, for example, Kenyan newspapers carried at least three reports of lake and river tragedies in tropical Africa as a result of boats and ferries capsizing. Seven people died on Lake Naivasha (Kenya), 19 on the Luapula River on the Zambia–Zaïre border and 88 on Lake Volta in Ghana. There was also a report of five deaths in a Nigerian rail accident.

Conclusions

In seeking to pull together the principal arguments and strands of this book, this chapter has highlighted some important current policy issues in relation to transport and development as the twentieth century and second millennium draw to a close. There are unlikely to be any radical departures from current trends, although these will intensify. Their impact and importance will by no means be similar everywhere, and I have sought to highlight how social, economic, political and territorial relationships impact upon, and in turn are impacted upon by, changing transport patterns, technologies and policies.

Perhaps the most encouraging trend of late has been the growing awareness of the interrelationships between different aspects of development processes, and the unsustainability of many present pathways. Far from being merely a technical and financial subject best left to engineers and economists, transport is essential to everyone in the contemporary world, which is becoming increasingly closely integrated. To be sure, relative accessibility is as important as absolute accessibility, and the balance between different places within the same system is seldom constant for long. Particular groups of people living and working in specific places are being differentially affected.

One of the principal paradoxes of ongoing processes of globalisation and the introduction of more sophisticated transport and communications technologies is that systems of cities spanning the globe are coming far closer together in terms of interactions and ease of communications than many places within the same country. Professionals, diplomats, international agency staff and others with access to the resources and technologies required can communicate with their counterparts in almost any country instantaneously and can physically travel to other continents in a matter of hours. Conversely, however, it often still takes poor rural dwellers in the Amazon Basin, the Andean Altiplano (high Andes), the Sahel, Ethiopian Highlands, Himalayas, the hills straddling the Thai–Burmese border or the Highlands of Papua New Guinea, for example, several days to walk or paddle to the nearest bus route, administrative centre or health post. A person injured in an accident or suffering from a burst appendix in such areas is more likely to die (unless traditional remedies prove adequate or by chance an air ambulance service is accessible) than someone in a major city suffering from an acute and rare life-threatening condition, who can be airlifted to specialist treatment in another city or even country at short notice.

This sobering example highlights the nature and complexity of the development issues to which transport and the lack or inadequacy of it are central. These have been examined in different ways throughout this book in an effort to stimulate debate and critical thought about issues many people take for granted. They can be, and sometimes are, quite literally matters of life, its quality, and death.

Key ideas

1 Current trends towards greater use of intermodal and integrated transport systems within major urban areas, over long distances and internationally are likely to intensify.
2 Many groups of countries are making increasing efforts to promote intraregional trade and other forms of interaction by enhancing their shared transport and communications infrastructure.
3 There is substantial scope for refining policies aimed at liberalisation and privatisation in the transport sector, so as to be more locally appropriate and to maximise the benefits of moving away from rigid past systems of operation, management and planning.
4 The growing awareness of the need to promote transport safety and sustainability is strongest in relatively high-income developing countries where congestion and pollution are most acute, as well as in poor countries which are experiencing heavy traffic volumes and accident tolls on often inadequate infrastructure.

Review questions and further reading

Chapter 1

Review questions

1 Outline the main assumptions and values implicit in the conventional approach to transport problems and solutions.
2 Explain what is meant by 'development' and outline at least one recent attempt to measure it more appropriately.
3 Discuss how an understanding of the difference between patterns and processes is important to the analysis of transport issues.

Further reading

Owen, W. (1987) *Transportation and World Development*, London: Hutchinson.
United Nations Development Programme (published annually) *Human Development Report*, New York: Oxford University Press 1994.
World Bank (published annually) *World Development Report*, New York: Oxford University Press 1994.

Chapter 2

Review questions

1 Explain why the introduction of new transport technologies in countries of the South is seldom an unqualified blessing.
2 Outline the difference between absolute and relative measures of transport provision and their respective uses.

3 Discuss to what extent different trends in the provision of road and rail networks are evident in countries of the North and South respectively.
4 Explain why non-motorised forms of transport are often regarded as inferior and obsolete in Third World cities.
5 Outline recent trends in worldwide civil air travel and freight transport, and account for the differential experiences of Africa and Asia in particular.

Further reading

Dick, H. W. and Rimmer, P. J. (1986) 'Urban public transport in Southeast Asia: a case study of technological imperialism?' *International Journal of Transport Economics* 13(2): 177–96.
Lea, J. P. (1988) *Tourism and Development in the Third World*, London: Routledge.
Owen, W. (1987) *Transportation and World Development*, London: Hutchinson.
Pirie, G. H. (1982) 'The decivilizing rails: railways and underdevelopment in southern Africa', *Tijdschrift voor Economische en Sociale Geografie* 73(4): 22–8.

Chapter 3

Review questions

1 How did the expansion of railway networks contribute to underdevelopment in many Third World countries?
2 Outline the principal elements of modernisation theory with respect to Third World development.
3 Explain how modernisation theory has been applied to the development of transport networks and transport technological innovation in the Third World.
4 Evaluate the assertion that transport is a necessary but insufficient condition for development to occur.
5 Discuss how the transport infrastructure inherited by newly independent states affected the ability of the incoming governments to implement national development policies.
6 To what extent do you think that postmodernism and postdevelopmentalism have relevance for transport policy and planning in present-day Third World contexts?

Further reading

Blaut, J. M. (1993) *The Colonizer's Model of the World: geographical diffusionism and Eurocentric history*, New York: Guilford.

Ezeife, P. C. and Bolade, A. T. (1984) 'The development of the Nigerian transport system', *Transport Reviews* 4(4): 305–30.

Hoyle, B. S. and Smith, J. (1992) 'Transport and development' in Hoyle, B. S. and Knowles, R.D. (eds) *Modern Transport Geography*, London: Belhaven/ Wiley.

McCall, M. K. (1977) 'Political economy and rural transport: an appraisal of western misconceptions', *Antipode* 9(2): 98–110.

Pirie, G. H. (1982) 'The decivilizing rails: railways and underdevelopment in southern Africa', *Tijdschrift voor Economische en Sociale Geografie* 73(4): 22–8.

Slater, D. (1975) 'Underdevelopment and spatial inequality: approaches to the problems of regional planning in the Third World', *Progress in Planning* 4(2): 97–167.

Taaffe, E. J., Morrill, R. L. and Gould, P. R. (1963) 'Transport expansion in underdeveloped countries: a comparative analysis', *Geographical Review* 53(4): 503–29.

Chapter 4

Review questions

1 Discuss the claim that people lack accessibility and mobility because they are poor.

2 Why might the construction of a new road into a particular area not have immediate positive developmental effects?

3 Explain the concept of appropriate transport technology with respect to rural development.

4 Giving examples, discuss the possible impacts of new road construction on the health of local communities.

5 What is the rationale for including roads as one component of wider (integrated) rural development programmes?

6 Explain why so many of the effects of new roads are not foreseen or taken into account by transport economists and planners.

Further reading

Airey, A. (1989) 'The impact of road construction on hospital in-patient catchments in the Meru District of Kenya', *Social Science and Medicine* 29(1): 95–106.

Hathway, G. (1985) *Low-Cost Vehicles*, London: Intermediate Technology Publications.

Hine, J. L. (1982) *Road Planning for Rural Development in Developing Countries: A Review of Current Practice*, Laboratory Report 1046, Crowthorne, Berkshire: Transport and Road Research Laboratory.

Hine, J. L., Riverson, J. D. N. and Kwakye, E. A. (1983) *Accessibility and Agricultural Development in the Ashanti Region of Ghana*, Supplementary Report 791, Crowthorne, Berkshire: Transport and Road Research Laboratory.

McCall, M.K. (1977) 'Political economy and rural transport: a reappraisal of transportation impacts', *Antipode* 9: 56–67.

McCutcheon, R. (1988) 'The district roads programme in Botswana', *Habitat International* 12(1): 23–30.

Chapter 5

Review questions

1 Discuss the potential role of walking and cycling in major Third World cities.
2 Define paratransit, describe the variety of forms found in Third World cities and explain the conventional attitudes of urban authorities to these.
3 Analyse why the private motor car is widely regarded as a very mixed blessing by urban planners and managers in cities of the South.
4 Discuss the contention that, while public transport may be essential to the future efficient functioning of large Third World cities, an increasing proportion of urban trips are being made by private vehicles almost everywhere.
5 Evaluate the relative advantages and disadvantages of hi-tech mass rapid transit systems in Third World settings.
6 With respect to buses and/or paratransit, explain likely conflicts between transport liberalisation and privatisation on the one hand, and stability and safety within the industry on the other.

Further reading

Allport, R.J. (1984) 'Appropriate mass transit for developing countries', *Transport Reviews* 6(4): 365–84.

Armstrong-Wright, A. (1993) *Public Transport in Third World Cities*, TRL State of the Art Review 10, London: HMSO.

Barat, J. (1985) 'Integrated metropolitan transport: reconciling efficiency, equity and environmental improvement', *Third World Planning Review* 7(3): 241–61.

Barrett, R. (1990) *Urban Transport in West Africa*, World Bank Technical Paper 81, Urban Transport Series, Washington, DC: World Bank.

Darbéra, R. (1993) 'Deregulation of urban transport in Chile: what have we learned in the decade 1979–1989?', *Transport Reviews* 13(1): 45–59.

Dick, H. W. and Rimmer, P. J. (1986) 'Urban public transport in Southeast Asia: a case study of technological imperialism?', *International Journal of Transport Economics* 13(2): 177–96.

Dimitriou, H. T. (ed.) (1990) *Transport Planning for Third World Cities*, London: Routledge.
────── (1992) *Urban Transport Planning: a developmental approach*, London: Routledge.
────── (1995) *A Developmental Approach to Urban Transport Planning*, Aldershot: Avebury.
Fouracre, P. R., Allport, R. J. and Thomson, J. M. (1990) *The Performance and Impact of Rail Mass Transit in Developing Countries*, Crowthorne: TRRL Research Report 278.
Fouracre, P. R. and Maunder, D. A. C. (1986) *A Comparison of Public Transport in Three Medium Sized Cities of India*, Crowthorne: TRRL Research Project 82.
Gallagher, R. (1992) *The Rickshaws of Bangladesh*, Dhaka: University Press.
Gardner, G., Rutter, J. and Kukn, F. (1994) *The Performance and Potential of Light Rail Transit in Developing Cities*, Crowthorne: TRL Project Report 69.
Heraty, M. J. (1980) *Public Transport in Kingston, Jamaica and its Relation to Low Income Households*, Crowthorne: TRRL Supplementary Report 546.
Kalabamu, F. T. (1987) 'Rickshaws and the traffic problems of Dhaka', *Habitat International* 11(2): 123–31.
Maunder, D. A. C. (1990) *The Impact of Bus Deregulatory Policy in Five African Cities*, Crowthorne: TRRL Research Report 294.
Maunder, D. A. C. and Mbara, T. C. (1995) *The Initial Effects of Introducing Commuter Omnibus Services in Harare, Zimbabwe*, Crowthorne: TRL Report 123.
Setchell, C. A. (1995) 'The growing environmental crisis in the world's megacities: the case of Bangkok', *Third World Planning Review* 17(1): 1–18.
Spencer, A. H. and Madhavan, S. (1989) 'The car in Southeast Asia', *Transportation Research A* 23A(6): 425–37.
Takyi, I. K. (1990) 'An evaluation of jitney systems in developing countries', *Transportation Quarterly* 44(1): 163–77.

Chapter 6

Review questions

1 Outline recent changes in civil aircraft technology and environmental regulations and explain the developmental impacts on countries of the South.

2 Discuss the paradox that Panama and Liberia, two small, poor countries, have among the world's largest registered shipping fleets.

3 Evaluate how the shift of passengers and freight from rail to road is affecting traffic and road conditions on major trunk routes.

4 Explain the importance of transport to the development of extended metropolitan regions such as Singapore.

Further reading

Hilling, D. (1990) 'African development and the maritime sector', *The Dock and Harbour Authority* 70: 330–5.

Mwase, N. (1987) 'Zambia, the TAZARA and the alternative outlets to the sea', *Transport Reviews* 7(3): 191–206.

Pirie, G. H. (1990) 'Aviation, apartheid and sanctions: air transport to and from South Africa, 1945–1989', *GeoJournal* 22(3): 231–40.

Robinson, R. (1989) 'The foreign buck: aid-reliant investment strategies in ASEAN port development', *Transportation Research A* 23A(6): 439–51.

Rodrigue, J.-P. (1994) 'Transportation and territorial development in the Singapore Extended Metropolitan Region', *Singapore Journal of Tropical Geography* 15(1): 56–74.

Chapter 7

Review questions

1 Explain the distinction between liberalisation and privatisation policies and how they have come to be implemented in countries of the South.

2 Discuss the problems experienced with liberalisation and privatisation policies in practice and how they might be made more locally appropriate.

3 Outline the principal issues in relation to transport safety and how it might be improved.

4 Explain how countries of the South might be able to exploit new export opportunities provided by integrated and multimodal transport.

5 Evaluate how and why 'sustainable thinking' can make a greater impact on transport policy and planning than it has done to date.

Further reading

Alokan, O. O. (1995) 'The road freight industry in Nigeria: new challenges in an era of structural adjustment', *Transport Reviews* 15(1): 27–41.

Anderson, D. (1989) 'Infrastructure pricing policies and the public revenue in African countries', *World Development* 17(4): 525–42.

Leinbach, T. R. (1989) 'Transport policies in conflict: deregulation, subsidies and regional development in Indonesia', *Transportation Research A*, 23A(6): 467–75.

Roschlau, M. W. (1989) 'Nationalisation or privatisation: policy and prospects for public transport in Southeast Asia', *Transportation Research A* 23A(6): 413–24.

Index

Printed in the United Kingdom
by Lightning Source UK Ltd.
116056UKS00001B/30